THE NEW RULES OF WAR

ALSO BY SEAN McFATE

The Modern Mercenary

NOVELS WITH BRET WITTER

Deep Black
Shadow War

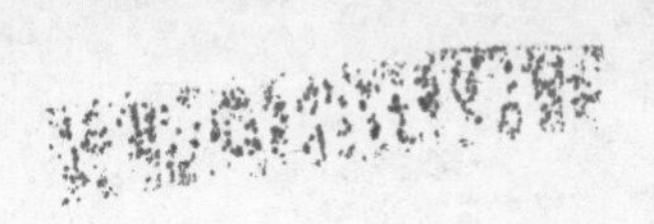

WILLIAM MORROW
An Imprint of HarperCollins*Publishers*

THE NEW RULES OF WAR

VICTORY IN THE AGE OF DURABLE DISORDER

SEAN McFATE

FOREWORD BY
GEN. STANLEY McCHRYSTAL (RET.)

The views expressed by this book are those of the author and do not reflect the official policy or position of the National Defense University, the Department of Defense, or the U.S. Government.

HarperCollins books may be purchased for educational, business, or sales promotional use. For information, please email the Special Markets Department at SPsales@harpercollins.com.

FIRST EDITION

Designed by William Ruoto

Library of Congress Cataloging-in-Publication Data

Names: McFate, Sean, author.
Title: The new rules of war : victory in the age of durable disorder / Sean McFate.
Description: First edition. | New York, NY : William Morrow, an Imprint of HarperCollins Publishers, [2019] | Includes bibliographical references.
Identifiers: LCCN 2018037595 | ISBN 9780062843586 (hardback) | ISBN 0062843583 (hardcover) | ISBN 9780062843609 (ebook) | ISBN 0062843605 (ebook)
Subjects: LCSH: War. | Military art and science. | BISAC: HISTORY / Military / Strategy. | TRUE CRIME / Espionage. | HISTORY / Modern / General.
Classification: LCC U105 .M38 2019 | DDC 355.02—dc23 LC record available at https://lccn.loc.gov/2018037595

ISBN 978-0-06-284358-6

19 20 21 22 23 RS/LSC 10 9 8 7 6 5 4 3 2 1

TO WARRIORS EVERYWHERE

For to win one hundred victories in one hundred battles is not the acme of skill.

To subdue the enemy without fighting is the acme of skill.

—SUN TZU

CONTENTS

FOREWORD

It is daunting to face an enemy whose singular goal is to destroy you. When that enemy's goal is chaos at any cost, the fight feels uniquely hopeless.

When I arrived in Iraq to lead the Joint Special Operations Command (JSOC) in 2003, I watched the nature of war change before my eyes. With a fleet of car bombs and zealous suicide attackers, al-Qaeda in Iraq (AQI) struck more civilian targets than any other terrorist group in history. Spaces that had been previously sacred—mosques, outdoor markets, and protected areas for religious pilgrims—were suddenly at the top of AQI's hit list.

Abu Musab al-Zarqawi was defeating us within the penumbras of new warfare, specifically his willingness to eschew any traditional rules or structure. AQI, less by design than organic mutation, was uncontained and unconstrained. While other terrorist organizations, including and perhaps especially wider al-Qaeda, had operated under strict policies and procedures, AQI thrived because they had no such shackles. Zarqawi aimed to wreak anarchic havoc on Iraq, no matter the method or price.

AQI never struggled to find supplies in Iraq's anarchy; in fact, mayhem fed the group's cache of resources. With the decision to disband the Iraqi Army and Baath Party, the United States had inadvertently created a supercenter for insurgents. The pool of angry and out-of-work Iraqis, abundance of weaponry flowing into the region, and revenue from its vicious kidnapping ransoms meant that Zarqawi and his men had almost everything they needed to succeed.

The final ingredient, though, was AQI's skill in leveraging Information Technology (IT) to its advantage. Jihadists could admire and contribute to AQI's efforts from far beyond Iraq's borders. Most importantly, though, IT allowed the group to control both the pace and narrative of violence. With the ability to connect its nodes at a rapid pace, IT facilitated AQI's growth into a broader network, which in turn fueled its ability to seem larger and speedier than it actually was.

Despite AQI's embrace of the new rules of war, at the outset I had confidence in JSOC's ability to adapt to these unconventional methods; this tactical flexibility was, in my mind, our specialty. And as in many wars against such foes, my troops won most firefights. We were better armed, better managed, and better trained.

However, our tactical successes gave both soldiers and policymakers the false impression that our strategy was working. In reality, though, we were simply carrying out discrete missions that were often brutally effective against our foe, but which were not truly rooted in any unifying national strategy or ultimate endgame. We were living one operation at a time; we celebrated our successes, but lacked wide enough perspective to clearly assess the impact we were having. And as veterans racked up tours, I realized that we had not invested enough effort in diagnosing the nuanced conditions which made AQI so resilient.

The more I pored over our situation, the clearer the solution became—however surprising it was. Aspects of JSOC which had previously made us so unrivaled—our structure, equipment, doctrine, and culture—were the very things constraining us. We were trapped in a cage of our own making; we believed ourselves to be tactically flexible, so much so that we stopped questioning whether our actions, or the nation's broader strategy were correct. In uncharted waters, my team and I endeavored to reimagine both JSOC's role in the war effort and its place in American foreign policy.

Our saga in Iraq spotlights a larger problem endemic not only to JSOC, but to the entire Western world—our culture does not force leaders to reckon with the intersection of strategy and adaptability. This is, in part, due to our incredible privilege. AQI had to constantly recalibrate simply to remain alive.[1] While America has absolutely faced terror and trauma, we remain a global superpower. We have, for too long, expected the world to play by our rules. In so doing, we failed to ask ourselves what would happen if those rules were incompatible with reality.

Paradoxically, America seems to remain fearful of strategic adaptability in any setting. We are wedded to the notion that we shouldn't change a policy until it has failed, unwilling to ask ourselves how we can do better. Clinging to the status quo is, in the short-term, an easy course of action, but it is also a dangerous one.

The world is changing faster now than ever before, and unsurprisingly, new styles of leadership will become more important

---

[1] It is worthwhile to note that AQI continues to adapt and evolve today, but the group goes by a different name now—the Islamic State.

than ever. We can no longer rely on the flexible iconoclast or the by-the-book manager alone—we must combine outside-the-box and ordered thinking. This kind of hybrid leadership will be necessary not only for success in warfare, but in other worlds as well.

Leaders must seek to prevent crises, not simply wait for them to happen. As we learned in Iraq, consciously sacrificing long-term strategy for short-term certainty is both unwise and unsustainable. We were lucky that, despite being on our heels, JSOC was able to withstand so many blows before we recognized the need to reinvent ourselves. Some organizations aren't so lucky; just look at the slew of businesses who were slow to react to the rise of Amazon. However, the question remains: how do we create these strategically adaptable leaders in a world afraid of change?

The first step is to identify our cultural problem, as we did in Iraq, and as Sean McFate expertly does in this book. As he so aptly reports, military leaders must combine a level of elasticity and big-picture thinking when confronted with new styles of conflict. Accordingly, we must come to terms with the fact that following yesterday's rules of war will not lead to today's (or tomorrow's) successes—that awareness alone can save lives. We must begin to grapple with the consequences of the new rules of war; if not, we will all be left behind.

—Gen. Stanley McChrystal (Ret.)

AUTHOR'S NOTE

For readers interested in more information about specific references and sources in the text, please see the notes and selected bibliography at the end of the book.

STRATEGIC ATROPHY

Why has America stopped winning wars?

On June 5, 1944, the day before D-day, General George S. Patton strode onto a makeshift stage in southern England to address thousands of American soldiers. "Americans play to win all of the time," said Patton. "That's why Americans have never lost nor will ever lose a war, for the very idea of losing is hateful to an American."

Since then, the American military has experienced nothing but loss. Korea is an ongoing stalemate. Vietnam turned Communist. The wars in Iraq and Afghanistan have failed, too. ISIS destroyed vast swaths of Iraq, and Iran has its tentacles in Baghdad. The Taliban controls more of Afghanistan than the local government does. Wars since 1945 have squandered American blood, wasted trillions of tax dollars, and damaged national honor, while resolving nothing on the ground.

We are losing. People are worried. Those not yet convinced that we are failing may owe their conviction to a false concept of victory. Winning is not about who killed more enemies or seized more territory. Those factors are irrelevant. The only thing that matters is where you are when the war is over. Did you achieve the objectives you set at the beginning? If the answer is no, then you can't claim victory. Some people try to cheat by rationalizing failure or redefining objectives, but history is never fooled. The last time the United States won a conflict decisively, the world's electronics ran on vacuum tubes.

This problem is not one of Democrats versus Republicans; rather, it's an American one. Presidents of both parties have either led us into wars we cannot win, or failed to get us out of wars as promised. But don't blame the White House for everything—Congress has been AWOL since the Truman administration. The last time it officially declared war was World War II, despite armed conflicts in Korea, Vietnam, Grenada, Panama, Somalia, the Balkans, Iraq (twice), Afghanistan, and Syria. What exactly did our soldiers die for? As a former troop leader, I want to know. I'm sure I'm not alone.

This isn't just happening to the United States. Over the last seventy years, a disturbing trend has emerged: the West has forgotten how to win wars. It's obvious, but no one talks about it, because the implications are too terrifying. The United Kingdom and other Western powers have struggled in their conflicts since World War II, from the French in Indochina to NATO in Afghanistan. The West is stuck in quagmires everywhere. The UN's peacekeeping missions have fared no better. Modern war's only constant is that the world's strongest militaries now routinely lose to their weaker enemies.

The West has the best troops, training, technology, equipment, and resources—so what's the problem?

Some experts think Western countries should double down on one or more of their core military strengths, such as whiz-bang technology or billion-dollar budgets, but we have been doing that for decades and nothing has improved. This solution is representative of the classic definition of insanity: doing the same thing over and over and expecting different results. Since World War II, destitute, untrained, low-tech militias armed with primitive weaponry have foiled military juggernauts routinely. France was defeated in Algeria and Indochina, Great Britain in Palestine and Cyprus, the USSR in Afghanistan, Israel in Lebanon, and the United States in Vietnam, Somalia, Iraq, and Afghanistan. Waging war the same way in the future is not the answer.

Others are in denial. A few people still believe we have won, or at least have not lost, in places like Afghanistan. They stand alone. Several polls conducted by the Pew Research Center in 2018 found that over half of all Americans think the wars in Iraq and Afghanistan have "mostly failed." Polls in the United Kingdom confirm the same grim conclusion. Even Senator John McCain conceded that the war in Iraq, which he fought so hard to launch and escalate, "can't be judged as anything other than a mistake, a very serious one, and I have to accept my share of the blame for it."

Some place their faith in the United Nations and international law to resolve armed conflicts. These are the dreamers. They see war as they wish it to be, not as it is. The United Nations did nothing to stop the genocides in Rwanda and Darfur. Nor did it challenge Russia's theft of Crimea or curtail decades of slaughter in the Middle East. The favorite weapon of such dreamers is the strongly worded memo, which explains much. The Law of Armed Conflict is equally charming but ineffectual. As any veteran will tell you, the laws of war are a marvelous fiction. These "laws" exist in name only. No one can legislate combat, or regulate it, and it is hubris to try. Kindhearted solutions to war just get more people killed.

Others throw up their hands in frustration, saying that war is simply too chaotic to comprehend, so why should we try? These are the quitters, and they are wrong. Oddly, many of them are foreign policy experts who have tried and failed to put forth sound strategies, so they submit that it cannot be done. These experts once assured us that Saddam Hussein had weapons of mass destruction, and that the wars in Iraq and Afghanistan could be won quickly and cheaply. When those things did not happen, they told us we needed to nation-build in order to win. Then they promised that a counterinsurgency strategy would fix everything. When it did not, a "surge" of troops would surely save both wars. That failed, too, and now Iraq and Afghanistan are worse off than before the United States arrived.

Experts walk away, saying war is unknowable. However, it's unknowable only to them. These same experts will insist there are no "rules" of war, but don't believe them. I am always astonished by this ignorance, since it suggests that armed conflict can never be understood. The truth is, humans have been successfully studying warfare for thousands of years, from Sun Tzu in ancient China to Carl von Clausewitz in nineteenth-century Europe. We still read the masters today. They show us that war is knowable, and there are timeless ideas on how to win. Call them ideas, principles, rules—it's just semantics. People like to argue semantics when they lack ideas of their own.

So why does the West continue to lose wars, even to vastly inferior foes? The problem is not the troops or resources—the West has the best. It's the way we think. The problem is our strategy. We lose because of our own strategic incompetence.

One of our most serious obstacles today is that we do not know what war is, and if we do not understand it, then we cannot win it. The French historian Marc Bloch witnessed the German blitzkrieg crush the French military in 1940, and he lamented how "our leaders . . . were incapable of thinking in terms of new war. . . . [Their] minds were too inelastic."

Minds are too inelastic today. Western militaries have become paradigm prisoners of something called "conventional war" strategy. It's modeled on World War II, but it has devolved into this: Deliver munitions into the enemy, who absorb it passively and then retreat home. Whoever kills more enemy troops and captures the most territory wins. This is victory. But in reality, it's a ticket on the *Titanic*. It will always fail, because the enemy, too, gets a vote, and no one fights "conventionally" anymore—except us.

The West is losing because it suffers from strategic atrophy. We yearn to fight conventional wars like it's 1945, our glory days, and then we wonder why we have stopped winning. War has moved on, and our enemies have moved on with it. But we are stuck in the fantasy of yesteryear, and that's why we are failing. We do not know how to fight other kinds of wars, especially the confusing, endless ones of today. Rather than face the future, experts turn to the past and imagine robot wars and grand air-naval battles against China that resemble World War II with better technology.

Forget what you know—wars of the future will look nothing like those of the past. If there will be a major conflict between great powers, such as between the United States and China, why do people always assume it will be fought conventionally? It won't.

Conventional war is dead. Those stuck in the traditional mind-set will probably not even recognize future conflicts as wars at all, until it is too late.

There is more to war than warfare, and more to warfare than killing. Understanding this is the key to winning. Modern conflict is governed by new rules, ones our enemies have grasped but we have not. Soon our adversaries will surpass us, and we will suffer a big defeat. Ancient Rome thought itself unassailable until it was sacked by Visigoths in AD 410. The modern Western world is no different. Nothing lasts forever, and barbarians are nearing the gate. Even an undefeated army can lose a war.

Losing is hateful to me, as it is for all Americans. As a veteran, I'm sickened to see friends killed in action owing to our leadership's low strategic IQ. As a taxpayer, I'm disgusted that our government has blown trillions of dollars abroad and only made the situation worse on the ground. As an American, I loathe seeing our national honor tarnished by low-level enemies. This is not what past generations sacrificed for, and the United States deserves better. So does the world.

The New Rules of War will help remedy the West's strategic atrophy. Some rules are ancient, others are new, and all are powerful. Observe them and they can deliver victory. People who claim they know how to win future wars are usually wrong. This book is different. Undergirded by scholarship and real-world experience, it identifies trends that have existed for the past seventy years and will continue into the next seventy. Its descriptions look like the future only because we are too accustomed to embracing the past, at least when it comes to war.

War is one of humanity's constants. No matter how enlightened we become, we'll still spend our time killing one another. As such, it is inevitable that today's younger generation will experience war. The only question is when. In the future, some conflicts will be regional, while others will affect us all. Some will be small, others will be big. All will be horrifying.

The good news is that we can still win. War is knowable, and half of winning is knowing what it looks like. The bad news is we have forgotten how. Western strategic thought is antiquated and incapable of safeguarding us. Many think the biggest threats today are terrorists, rogue states, and revisionist powers like Russia and China. While these opponents are bad, they are not the worst. Conventional strategists can see only one country or group at a time, but the more important challenges are systemic. Worldwide volatility is getting to the point that chaos threatens. If we are to endure, we must learn how to win in an age of disorder.

DURABLE DISORDER

The twenty-first century is maturing into a world mired in perpetual chaos, with no way to contain it. What has been tried so far has failed, making conflict the motif of our time. People intuitively know this, but here are some arresting facts.

The number of armed conflicts has doubled since World War II, and research shows that Americans were substantially safer in the Cold War years than they are today. Of approximately 194 countries in the world, nearly half are experiencing some form of war. The phrases "peaceful resolution" and "political solution" have become punchlines. Studies reveal that 50 percent of peace agreements fail in five years, and that wars no longer end unless one side is obliterated, like the Tamil Tigers in Sri Lanka or the Chechens in Grozny. Instead, modern conflicts smolder in perpetuity without a clear winner or loser.

Ancient rifts, such as that between Sunni and Shia Muslims, reopen and destabilize entire regions. UN peacekeeping fails, mostly because there is no peace to keep. Nothing seems to work: high-stakes negotiations, superpower interventions, track II diplomacy, strategic nonviolence, nation-building, or winning hearts and minds. Everything fails. Conflicts breed like tribbles, and the international community is proven powerless to stop them.

This growing entropy signifies the emergence of a new global system that I call "durable disorder," which contains rather than solves problems. This condition will define the coming age. The world will not collapse into anarchy; however, the rules-based order we know will crumble and be replaced by something more organic and wild. Disorder has taken over the Middle East and Africa, significant portions of Asia and Latin America, and is creeping into Europe. Soon it may be in North America.

The defining feature of durable disorder is persistent armed conflict, but not as you know it. We must ask—and answer—unsettling questions about modern and future wars: Who will fight and why? How will they fight? How will we win?

In the coming decades, we will see wars without states, and countries will become prizes to be won by more powerful global actors. Many nation-states will exist in name only, as some practically already do. Wars will be fought mostly in the shadows by covert means, and plausible deniability will prove more effective than firepower in an information age.

If there are traditional battles, they will not prove decisive. Winning will change, and victory will be achieved not on the battlefield but elsewhere. Conflicts will not start and stop, but will grind on in "forever wars." Terms like "war" and "peace" will come to mean nothing. The laws of war will fade from memory, as will the United Nations, which will prove useless in the face of conflict. If it persists, it will be only as letterhead stationary.

Mercenaries will return, not slinging AK-47s but flying drone gunships and auctioning special operation forces teams to the highest bidder. Some may take over countries, ruling as kings. Privatizing war changes warfare in profound ways, a fact incomprehensible to traditional strategists. It also warps international relations. When the super-rich can rent militaries, they become a new kind of superpower, one capable of challenging states and their rules-based order. Big oil companies will have private armies, as will random billionaires. In fact, this is already happening. Drug lords possess private forces and take over countries, turning them into zombielike "narco-states."

The most effective weapons will not fire bullets, and nonkinetic elements like information, refugees, ideology, and time will be weaponized. Big militaries and supertechnology will prove inept. Nuclear weapons will be seen as big bombs, and limited nuclear war will become acceptable to some. Why do we assume the nuclear taboo will last forever?

Others are already fighting in this new environment and winning. Russia, China, Iran, terrorist organizations, and drug cartels exploit durable disorder for victory, hastened by the West's strategic atrophy. These foes have significantly fewer resources than the West but are more effective in warfare.

Why? Because they are playing by new rules that we have not yet recognized.

We are dangerously unprepared because war has moved on, yet the United States and other Western powers have not. They assume the future will look like the past, and that traditional strategies will work in the decades to come. Should this foolishness continue, we will eventually be tested, and we will fail. However, this can be averted if we act now, before the crisis.

The New Rules of War will make the conventional warrior's head explode, but that's expected. They work because they embrace the essence of war for what it is, not as some wish it to be. Only by following these rules can we prevail in an age of durable disorder. If we do not, terrorists, rogue states, and others who do not fight conventionally will inherit the world.

WHY DO WE GET WAR WRONG?

Alice in Wonderland is a magnificent guide for understanding strategic atrophy. To paraphrase one teaching: If you don't know where you are going, any road will get you there. When the Cheshire Cat said this to Alice, he could have been discussing military operations today. Let's face it, we're lost. Why do we continue to get war so wrong, decade after decade? Where do all those bad ideas come from?

War futurists. These are the people who dream up future war scenarios. They fill our heads with the make-believe battles of tomorrow that drive strategic decisions today. Their assumptions shape Washington's worldview about what's coming to kill us and how we should fight back. In other words, their ideas shape our current concept of war and victory. The problem is, they get it wrong—all the time. Lawrence Freedman, a preeminent war scholar, surveyed modern history and found that predictions about future war were almost always incorrect.

War futurists exemplify the Washington adage "No stupidity before its time." There's zero accountability for futurists who are consistently outsmarted by a Magic 8-Ball. Nonetheless, Washington continues to listen to them, and this influences military spending and strategy. Militaries invest billions of dollars preparing for the next war, but what should they buy? It all depends on your vision of future war, and it matters when aircraft carriers cost $13 billion each (before adding aircraft and crew). How many carriers will you need in the future? One, ten, more? What strategies does an aircraft carrier enable or shut out? What are we not buying that we may need instead? Choose wisely, because people pay with their lives, and the nation may be destroyed if you're wrong.

The edifice of military planning is built on the assumptions of futurists, so when they get it wrong—and they almost always do—everything downstream goes awry, too. After keeping this up for decades, countries stop winning wars. Some are even destroyed. To reverse this trend, we must jettison these charlatans. They deceive us and foster strategic atrophy. To filter them out, you need to know who they are and what they are saying. It may surprise you.

FALSE PROPHETS

Who are the influential war futurists? One would assume that generals, intelligence officers, university faculty members, and think tank fellows would be the thought leaders. However, these individuals are eclipsed by people who have the imagination to steer pop culture. The most influential war futurists in the West are artists, novelists, and filmmakers; people whose visions inspire us all. But they also delude us. What makes a good movie does not make an effective strategy, and vice versa. Despite this, Washington's visions of future war look like they came off a Hollywood set.

War futurists come in three flavors: nihilists, patriots, and technophiles. Nihilists work in the zombie and postapocalyptic modes, as exemplified by novels like *World War Z* and the Mad Max films, which showcase a Hobbesian future of ruination. These works are entertaining and horrifying in equal measure. The future they portray is not disorder but hell, and they offer little insight beyond lone survival.

The second group, the patriots, creates thrillers in the mold of Tom Clancy. For them, the flag is front and center, and technology is a character in the story of war. They glorify the military's vision of itself: industrial-strength conventional-war firepower with all the trimmings. To them, the future of war looks like World War II with better technology.

For example, take Clancy's bestselling novel *Red Storm Rising*, which was written in the mid-1980s. It portrays a World War III in which the Soviets draw first blood through the use of dishonorable tactics. In the end, the good guys (NATO) win decisively, with America leading the charge in a conventional fight. Nukes are conspicuously absent, making it more like World War II than III, but readers didn't mind this improbable omission. When I was an army cadet, I remember officers walking around with this novel as if it were a strategic oracle. They still do this, with novels by Clancy's successors.

Clancy got everything wrong. Only three years after *Red Storm Rising*'s publication, the Berlin Wall fell, and the Soviet Union with it. In reality, the USSR was never a threat to the West in the 1980s, not even close. That military officers thought *Red Storm Rising* prescient displays how clueless they were about the enemy. Then again, the CIA missed it, too—"missed by a mile" according to one former CIA director—in one of the biggest intelligence failures in history. Perhaps they were reading Clancy instead of intel reports, as the agency depicted an expansionist, invulnerable Soviet Union in its briefings from the late 1980s. But how could the CIA have been truly surprised, when its primary mission was to forecast the strength of the USSR? Cheerleader in chief for the Clancy version of the USSR was Robert Gates, who went on to become the head of the CIA and the US secretary of defense despite his blunder. Sometimes there is little accountability in government, which presents another challenge when preparing for future wars.

Members of the third group, the technophiles, are the most deceiving war futurists. They fetishize exotic machinery and foretell the rise of Terminator robots, the creation of Iron Man suits, and the dawn of some version of Skynet or the Matrix. Shockingly, the Pentagon buys it. Literally. It has spent $80 million on an Iron Man suit called TALOS (Tactical Assault Light Operator Suit), a powered exoskeleton that the military hopes will be bulletproof and weaponized, and also able to monitor the wearer's vitals and give him increased strength and perception. The only problem is that it needs Tony Stark's Arc Reactor to power it, a figment of Marvel Comics' imagination. Undeterred—and perhaps to trick Congress into spending more money—the military hired Legacy Effects, a Hollywood costume-design firm, to build a fake rendition of the suit.

Technophiles use scare tactics to browbeat audiences into submission, and one of their favorite scares is the "robot revolution." This scenario envisions a future in which machines become sentient, rise up, and annihilate the human race. Seen that movie before? Some futurists believe that if we are lucky, robots will keep a few of us as pets. The Pentagon even coined a buzz phrase for the rise of autonomous killer robots, the Terminator Conundrum, which was directly inspired by the blockbuster movie franchise.

However, studies show that artificial intelligence (AI) can currently barely accomplish basic cognitive tasks. One Stanford University experiment involved feeding pictures of objects into a machine-learning algorithm, so it could learn to identify them. When finished, it made goofs even a child would not make. For instance, one picture showed a baby clutching a toothbrush, which the machine labeled "A young boy is holding a baseball bat." The rise of the machines will not happen anytime soon.

Cyberwar doomsayers are the biggest con artists among technophiles—a genuine achievement. Despite what experts claim, nobody really knows what "cyber" means, other than ones and zeroes in space. So far, no one has been killed by a cyberweapon, but such facts do not stop these futurists from dreaming up Armageddon scenarios in which a single hacker shuts down New York City, the Eastern Seaboard, or the planet. They draw inspiration from movies like *Skyfall*, a James Bond film in which every time someone touches a keyboard, that person becomes a god. But as any real hacker will tell you, hacking is boring and makes for bad TV.

Hollywood's hyperbolic portrayals of cyberwar influence Washington at the highest levels. Back in 2011, Leon Panetta, then the CIA director, warned Congress: "The next Pearl Harbor could very well be a cyberattack." In 2017, the Department of Energy declared that America's power grid "faces imminent danger" of a cyberattack that could produce nationwide blackouts, causing billions in damage and threatening lives. Such alarmist warnings are reckless hokum. Research shows that squirrels pose a greater threat than hackers when it comes to blackouts. Perhaps the CIA should include RodentWar on its list of "next Pearl Harbors."

Cyberwar is magical thinking. However, cyber experts demur and showcase Stuxnet as proof that cybertechnology is not just a new weapon of war, but a new way of war. Stuxnet was an American-Israeli computer worm injected into Iran's nuclear facility network at Natanz in 2010. The worm took control of some computers and ordered the nuclear centrifuges to spin apart, reportedly destroying a fifth of them. Many asserted (without evidence) that this caused significant damage to Iran's nuclear weapons program, and everyone else weirdly believed this. A much-read *Vanity Fair* article claimed the episode represented the future of war, declaring: "Stuxnet is the Hiroshima of cyber-war." In reality, Stuxnet had no effect on the Iranian nuclear program. It did not destroy it or even meaningfully delay it. The Iranians simply replaced the broken centrifuges, ran an antivirus program, and went back to developing nuclear weapons. Stuxnet is pure hype.

Cyber is important, but not in ways people think. It gives us new ways of doing old things: sabotage, theft, propaganda, deceit, and espionage. None of this is new. Cyberwar's real power in modern warfare is influence, not sabotage. Using the internet to change people's minds is more powerful than blowing up a server, and there's nothing new about propaganda.

If there is one lesson from the past seventy years of armed conflict, it's this: technology is not decisive in modern war. Technophiles remain inexplicably oblivious to this fact. Since World War II, high-tech militaries have been routinely stymied by luddites: France in Indochina and Algeria; Great Britain in Aden, Palestine, and Cyprus; the USSR in Afghanistan; Israel in Lebanon;

the United States in Vietnam, Iraq, and Afghanistan. Sexy technology does not win wars.

Washington draws more insight about future wars from Hollywood and hacks than from the study of war itself. No wonder we are not winning. To prepare for tomorrow, we must embrace war as it is, not as we wish it to be. Owing to this, some experts reject "futurism" altogether, on the basis that it is unknowable. This is the quitter crowd again. The best we can hope for, such individuals infer, is a lightning response to whatever the enemy hurls at us. This is a reactionary strategy, which is the same as no strategy.

We can do better. Genuine war futurists do exist, but they are rare.

TRUE PROPHET

General William "Billy" Mitchell did not suffer fools gladly. An American war hero and pilot during World War I, he had seen the future, and it was air power. He knew the world must not be deluded into the belief that "the war to end all wars" had really achieved that outcome. "If a nation ambitious for universal conquest gets off to a 'flying start' in a war of the future," he said, "it may be able to control the whole world more easily than a nation has controlled a continent in the past."

After the war, Mitchell proselytized for the importance of air power. His forecast was heresy to the conventional war thinkers of his day. He even claimed an airplane could sink a battleship. Back then, in the era of the superdreadnought, aircraft were little more than motorized kites. Mitchell was laughed out of the room.

Undeterred, he suggested that aircraft carriers should replace battleships, the long-reigning kings of the sea. In 1924, when he said this, only a stunt pilot would have considered landing a plane on a moving ship.

General John J. Pershing, perhaps to keep Mitchell out of further trouble, sent him on an inspection tour of the Pacific. Months later, he returned with a 525-page report, predicting war between Japan and the United States, initiated by a Japanese surprise attack from the air. Incredibly, he also predicted it would occur at Pearl Harbor. "Japan knows full well that the United States will probably enter the next war with the methods and weapons of the former war," he wrote. "It also knows full well that the defense of the Hawaiian group is based on the island of Oahu and not on the defense of the whole group." In other words, the Japanese needed to hit only one island to cripple America's Pacific fleet.

The top brass already thought Mitchell was nuts, but this new assertion went too far. He was deemed insubordinate, a cardinal sin in the military. Some ideas, it seems, are too dangerous for consideration, especially when they regard the future of war.

His court-martial soon followed. Of the thirteen military judges, none had aviation experience; three were removed following the defense's challenges for bias, including Major General Charles P. Summerall, the president of the court. After thirty minutes of deliberation, Mitchell was found guilty and suspended from the army without pay for five years. In disgust, he resigned. For the next decade, he preached tirelessly about the coming age of air war, when battles would be fought in the sky over who would rule continents. Many thought him an entertaining fruitcake.

Weakened by his struggle, the old campaigner died a decade later, in 1936, at the age of fifty-six. He elected to be buried in Milwaukee, his hometown, rather than at Arlington National Cemetery.

Five years later, the Japanese attacked Pearl Harbor by surprise, with airplanes. Within two hours, they sunk or damaged eight American warships, including the famed USS *Arizona*, destroyed 188 aircraft, and killed 2,403 people. The navy claimed it had been caught "completely by surprise," and blamed an unscrupulous enemy for its failure. In reality, the 2,403 casualties were killed by groupthink as much as by Japanese bombs.

The United States and Japan went on to fight one of the biggest naval engagements in history, the Battle of Midway, entirely with aviation. Never did the two fleets see each other. When it was over, the aircraft carrier had supplanted the battleship as supreme on the ocean, just as Mitchell had foretold twenty years earlier.

Mitchell saw the future, but no one believed him. Years after he died, the military did own up to its error, in its own way. It named a bomber after him.

Billy Mitchell teaches us that changing strategic minds is difficult, especially when it comes to the future of war. The stakes are considerable and the dogma thick. People are not always ready to receive the future.

CASSANDRA'S CURSE

Many experts think that predicting the future of war is a loser's game. Robert M. Gates, the former US secretary of defense, used to quip that Washington's predictions about the future of war have been 100 percent right, zero percent of the time. Perhaps he was referring to his own dismal record as an intelligence analyst, since Billy Mitchell shows us that forecasting the future is possible. However, when a genuine war prophet speaks, no one listens. In Greek mythology, Cassandra was given the power of prophecy, but cursed that no one believed her. The curse of the war prophet is Cassandra's Curse.

Given Cassandra's Curse, how do you spot true war futurists? First, if they are taken seriously at all, they are vilified by the groupthink mob for challenging the establishment. Second, they are often scholar-practitioners. That is, they are intellectual warriors who have experienced war in a new way and thus see things that conventional warriors and civilians do not. Understanding war is like swimming: you cannot learn the breaststroke in a classroom. At some point you need to jump in the pool, thrash around, and inhale water before you master it. Academics who learn war in libraries can learn only so much. Practitioners who mistake their war stories for macro insight are no better. There are exceptions, of course, but the best war futurists are scholar-practitioners. Third, true futurists possess the gift of sight—an eerie clairvoyance into what is to come—and most scholar-practitioners lack this. Glimpses into war's future are produced by a blend of new war experiences and preternatural intuition.

True prophets exist, and Mitchell is not unique. Major-General John "Boney" Fuller was a British tank officer in World War I. Like Mitchell, he saw how a new weapon would change war even though his peers did not. To them, the tank was an infantry support vehicle, sort of like a mobile foxhole. Fuller saw something different. In 1928, he wrote about tanks and aircraft fighting together to invade a country quickly, followed by infantry mopping up afterward. He believed the surprise and speed of this lightning strike would rapidly seize key terrain and shock the enemy into submission. The British dismissed him as a crackpot, but the Germans did not. They read Fuller's books and created the *Blitzkrieg*, or "lightning war," strategy that conquered most of Europe at the beginning of World War II. Fuller's ideas still govern mechanized warfare today.

William J. Olson is a scholar-practitioner of a different sort. Imagine it's 1983. President Ronald Reagan brands the Soviet Union an "evil empire" and authorizes the largest military buildup since that of the Second World War. Tom Clancy is writing *The Hunt for Red October*, and the United States and the USSR nearly start a nuclear war over a NATO exercise in Germany called "Able Archer." At the height of this Cold War frenzy, Olson points to a different future. Brushing aside the US-USSR conflict, he predicts that the future will descend into Islamic terrorism, ethnic conflict, failed states, and a global insurgency against the West. All of this was unfathomable back then. Unlike his peers, he spoke fluent Farsi and had spent the 1970s traveling in Afghanistan and Iran. What he saw that his contemporaries missed was a "parallel international system" festering dangerously. Olson was lambasted as a kook. As he now recalls, "The Blob ate my lunch, and then ate me." The Blob is the groupthink of the Washington consensus. Now we know that Olson was correct. What he saw, decades before anyone else did, was the post-9/11 world.

General Eric Shinseki knows a thing or two about fighting insurgencies. As a young officer, he served two tours in Vietnam before his right foot was partially blown off by a land mine. Thirty years later, he was fighting guerillas once again, this time as the commander of forces in Bosnia and Herzegovina when it was bad. On the eve of the Iraq invasion in 2003, he was chief of staff of the US Army and the four-star general with the most experience combating insurgencies. He advised against the plan of Donald Rumsfeld, then the secretary of defense, to control Iraq with only 100,000 troops. Rather, Shinseki told Congress that "something in the order of several hundred thousand soldiers" would be required for postwar reconstruction.

The Blob's reaction was swift. Paul Wolfowitz, the deputy defense secretary and neocon attack dog, belittled Shinseki's estimate as "wildly off the mark," and the general was forced out within months. Both President Bush and Secretary Rumsfeld made a point of not attending Shinseki's retirement ceremony, a clear snub, and they pulled Peter Schoomaker, a retired four-star general, out of retirement to run the army. The message was sharp: toe the line or be disgraced. The top brass received it loud and clear. From then on, no general questioned the obviously failing wars in Iraq and Afghanistan. A few years later, Shinseki's words proved prophetic when Bush fired Rumsfeld and announced a "surge" of additional troops to Iraq after miscalculating the numbers needed to stem sectarian violence.

In contrast to these prescient voices, today's conversation in Washington about the future of war is moribund. Think tanks conduct future-war scenarios that mirror last season's TV shows, and one institution even promotes science fiction stories as research. PhDs who have never smelled gun smoke in battle pontificate about war. These self-proclaimed strategists imagine high-tech conventional wars between great powers like the United States and China, involving smart drones, stealth ships, killer robots, rail guns, and artificial intelligence. It's D-day run by machines, and this absurd thinking is going to get us killed. Tomorrow's wars will have more in common with Cormac McCarthy than Tom Clancy. I know because I have walked that road.

I have looked at war from many sides. I began as a paratrooper in the US Army's 82nd Airborne Division, one of World War II's most storied units. I left the service to become a private military contractor in Eastern Europe and Africa. While the West became laser focused on Afghanistan and Iraq, I was on war's outer rim, fighting in Africa. What I saw there is invisible to conventional warriors and points to the future.

War's future is not what most people expect. I'm a professor at Georgetown University and the National Defense University, the premiere US Department of Defense war college, where I teach strategy and the art of war to senior military officers from around the world. Hearing the perspectives of my international students has galvanized my own. Western ideas of warfare are incredibly limited. And in the emergent age of disorder, they are about to be eclipsed.

We are facing a Billy Mitchell moment. The West's national security establishments are all stuck in a traditional-war mentality, which is as obsolete now as the dreadnought was in Mitchell's time. Future wars will not be fought like past ones, so why continually invest in legacy weapons and strategies? At the outset of World War II, France felt secure behind its Maginot Line, a string of fortresses along its border with Germany. Yet France fell in forty-six days. It is true now and always has been: militaries must adapt or die.

It's the same today. We invest trillions of dollars in legacy items like fighter jets and submarines, which play minor roles in modern war. Meanwhile, special operations forces and other weapons that work in the real world remain underfunded and overdeployed. In fact, a single aircraft carrier costs more than all US special operations forces combined. An urgent rethinking of priorities is needed.

Durable disorder is here, and those who know how to fight in it will win. The West does not, and it is on the path to defeat. Our strategies and weapons are deadly—to us. We need to catch up to our enemies and learn to play by war's new rules to win. Or we will die.

RULE 1: CONVENTIONAL WAR IS DEAD

Imagine a war. Several sides are fighting, but it is not clear who is on what side. The combatants do not wear regular military uniforms, and many are foreigners. They fight in the name of religion but act like monsters. Worse, they fight for the same god, labeling their enemy "apostate" and reserving the cruelest punishments for disbelievers. Groups splinter and turn on one another. The conflict becomes a holy mess to outside observers, and some even conclude that the religion itself is evil.

Civilians are prey, and the laws of war are nonexistent. Whole communities are raped and looted. Fighters carve out independent states in god's name and extort people of their wealth. They govern through terror, committing horrible human rights atrocities: children are slaughtered, women rounded up as sex slaves, men tortured, burned alive, beheaded, defenestrated, or worse.

In one city, a religious leader orders his fighters to put all inhabitants to the sword. And they do. One witness recounts: "Everyone—women, old and young, and sick, and children and pregnant women were cut to pieces at the point of a dagger." Babies were "taken by the feet and dashed against walls." Thousands more flee into the countryside, only to die slowly of thirst. The international community screams outrage but does little to stop the massacres.

People flee the war zone, creating a tidal wave of refugees that floods other countries, destabilizing them. The region sinks into chaos. Other powers intervene, exploiting the situation for their own interests and waging proxy wars against enemies, but they, too, become mired in the tar pit of war. Humanitarians decry all sides, penning invectives and condemning the bloodbath, but achieve nothing. Meanwhile, the disorder feeds on itself, resulting in perpetual conflict with no resolution in sight.

Is this the Middle East today? No.

The War of Eight Saints took place in Italy from 1375 to 1378, but the parallels between it and the Middle East today are stunning. The religion in question is not Islam but Christianity. Instead of Sunni fighting Shia over "true Islam," papists fought anti-papists for the soul of the Catholic Church. Warriors did not wear standard uniforms, and many were foreigners. In the War of Eight Saints, most fighters were mercenaries hailing from every corner of Europe. They, too, professed to fight for or against the pope, but many were interested only in coin or adventure. The same could be said of jihadis today. Terrorists are masked and wear a collage of military fatigues. Sunni and Shia come from all over the Middle East and North Africa. Combatants in both wars were and are savage.

All fought for god but behaved like devils, damning the innocent to a living hell. In 2014, the terrorist group known most commonly in the West as the Islamic State of Iraq and al-Sham (ISIS) took the city of Sinjar, Iraq. They rounded up the inhabitants and slaughtered them in the name of Allah: men, women, and children. Five thousand were killed. Even more fled up Mount Sinjar outside the city, dying of thirst. The War of Eight Saints had its own Sinjar: the massacre of Cesena, a small city in northern Italy. In 1378, Cardinal Robert, the pope's envoy, ordered the mercenary captain John Hawkwood to kill all of the town's civilians—five thousand of them—as God's punishment. Hawkwood did. Tellingly, it did not hurt either man's career. Hawkwood became one of the most celebrated and wealthy mercenaries of his day, and his visage still adorns Florence's famed cathedral. Cardinal Robert later became a pope himself, known as antipope Clement VII during the papal schism. Some do well by war.

Both conflicts sucked entire regions into anarchy. Syria and Iraq remain the epicenter of an ancient feud between Sunni and Shia, one with no permanent resolution in sight. The War of Eight Saints was fought for three years, but that was only the beginning. It birthed the great papal schism that split the Catholic Church from 1378 to 1417, wreaking pandemonium across Europe. The fight between papists and anti-papists would continue through the Reformation and the Thirty Years' War two centuries later, and arguably beyond.

The War of Eight Saints could be mistaken for the Middle East today because both illustrate the timeless nature of war: organized violence that means to impose the will of one on another. It is brutal, bloody, and unfair, whether in 400 BCE, 1300 CE, or today. Some things change—weapons, tactics, technology, leadership, circumstances—but the nature of war does not. This is the difference between war and warfare. Warfare is how wars are fought, and it is always changing. But the nature of war never changes.

People today confuse war and warfare, and this leads to big problems. The War of Eight Saints shows us what war is, but what is warfare now? For the West, it's called "conventional war."

THE WESTERN WAY OF WAR

There is no such thing as conventional versus unconventional war—there is just war. "Conventional war" is actually a type of warfare, and it's how the West likes to fight. Sometimes western militaries call it "Big War." Think of Napoleon or the world wars: Great powers duking it out with their militaries as gladiators, and the fate of the world dangling in the balance. Only states are legitimately allowed to do battle, making war an exclusively interstate affair, fought with industrial-strength armies. Firepower is king, and battlefield victory everything. Honor matters, as do the laws of war, and citizens are expected to serve their country in uniform with patriotic zeal. It's why we say "Thank you for your service" to vets in the airport.

World War II is the West's model for armed conflict. My grandfather fought in the Battle of the Bulge and called it the "good war." Others say it was fought by the "greatest generation." Nearly seventy years on, the demand for Second World War movies appears unstoppable, the supply inexhaustible. Like a handsome man in uniform, these films never really go out of style. There are over six hundred unique World War II movies, and four more were released in 2017. That conflict remains iconic because it represents the last time the West won decisively, unlike today. The frustrations that followed, from Korea to Afghanistan, are either forgotten or dismissed as "quagmires."

World War II remains paradigmatic for experts, too, who view its style of warfare as timeless and universal. Generals describe it using normative language like "conventional war," "symmetrical war," and "regular war." (I like the term "conventional war," but they all mean the same thing.) So strong is this dogma that other forms of combat are labeled "unconventional," "asymmetrical," or "irregular." These are snubs. Military campaigns waged by armed nonstate actors are not privileged as war, but belittled as something lesser.

Conventional war is state-on-state fighting in which the primary instrument of power is brute force and battle determines everything. It's a military-centric vision of global politics, which is why militaries cling to it and remain smitten to its call. The high priest of conventional war theory is Carl von Clausewitz, a Prussian general from the Napoleonic era. A hagiography exists around the man, and his book *On War* is enshrined in Western militaries as a bible. When I teach this text to senior officers at the war college, the room grows silent with reverence. His ideas constitute the DNA of Western strategic thought, and a few of his concepts, such as the "fog of war," have even made it into popular culture.

There is just one problem with conventional war: no one fights this way anymore. There is nothing conventional about it, because war has moved on. Despite this problem, conventional war remains our model, and this is why the West continues to lose against weaker enemies who do not fight according to our preferences. To win, we must ditch our traditional way of fighting, because it's obsolete. It is neither timeless nor universal. On the contrary, conventional war has a beginning, middle, and end.

A VERY SHORT HISTORY OF CONVENTIONAL WAR

The story of conventional war and the nation-state is one. While they have no precise birthday, one could argue it is May 23, 1618. That morning someone was thrown out of a window in Prague. Three others followed. Miraculously, all four survived, despite the seventy-foot drop onto cobblestones.

This set into motion a chain of improbable events that led to the Thirty Years' War, one of the bloodiest in European history. Catholics and Protestants fought each other without mercy. The armies of Sweden, then a superpower, destroyed 2,000 castles, 18,000 villages, and 1,500 towns in Germany alone. Disease and famine were rampant, and tens of thousands of people became refugees, wandering the plains of Europe and getting picked off by bands of roving mercenaries. Rape was routine. By the war's end, eight million were dead and most of central Europe was wiped out. The continent took a century to recover.

Out of this inferno came the Peace of Westphalia in 1648, which gave birth to a new international order, one uniquely ruled by states. Prior to that moment, Europe was a medieval free-for-all. Anyone with money could wage war, and everyone did. Kings, aristocratic families, cities, and even popes hired mercenary armies to do their bidding, no matter how petty. War was everywhere, all the time, and so was human suffering. It was the Wild West.

The "Westphalian Order" made states the sheriff, outlawing mercenaries and those who hired them. States invested in their own standing armies and began their ascendancy. The relationship among force, power, and world order is stark. Those who control the means of violence get to make the rules that others must obey—or die. Nonstate rivals became defenseless without mercenaries and were easily defeated. Old medieval powerhouses such as the church had no choice but to kowtow to state rulers. Soon, nation-states reigned above all others.

The Westphalian Order is the state-centric international system of today, the so-called rules-based order. It has many features, but the most important one is this: only states are sovereign, and everyone else is subordinate. States guarantee their sovereignty by clobbering anyone who might oppose them with their national militaries. Although gradual and imperfect, the Westphalian Order established modern diplomacy, international law, and the world we inhabit today.

The second-most-important feature of the Westphalian Order is armed conflict. Under it, only states were allowed to have militaries and wage war, enabling them to rule unchallenged. This made warfare an exclusively state-on-state affair, fought by national armed forces according to certain customs such as not killing prisoners or waving a white cloth to signal surrender. Later, these battlefield traditions were codified into the "laws of war," using instruments like the Hague and Geneva Conventions, which address only interstate conflict. All other forms of war were outlawed and viewed as illegitimate.

The Westphalian way of war became "conventional" in the mind of the age, and we are its inheritors. It was the only kind of warfare

Clausewitz knew, and it is what we teach today. Napoleon and the world wars remain paradigmatic conventional wars, as they were fought with national armies for flag rather than with mercenary armies for kings or popes. Battlefield victory determined winners and losers. These wars mirror the nation-state's own rise to glory; they are a recent invention, with most of history governed by empires, kingdoms and city-states. Both states and their way of war spread across the globe through European colonization, and today we have internalized them as timeless and universal, even though they are less than four hundred years old.

But the Westphalian Order is dying.

Today states are receding everywhere, a sure sign of disorder. From the weakening European Union to the raging Middle East, states are breaking down into regimes or are manifestly failing. They are being replaced by other things, such as networks, caliphates, narco-states, warlord kingdoms, corporatocracies, and wastelands. Syria and Iraq may never be viable states again, at least not in the traditional sense. The Fragile States Index, an annual ranking of 178 countries that measures state weakness using social science methods, warned in 2017 that 70 percent of the world's countries were "fragile." This trend continues to worsen.

As states retreat, the vacuum of authority has bred endless war and suffering, harkening a return to the Middle Ages in some parts of the world. These wars are not fought conventionally. Terrorism, ethnic cleansing, and other forms of violence by nonstate actors have eclipsed conventional interstate wars. The ability of the United Nations or the West to police the situation fades each year, while nonstate actors grow more powerful. International relations are returning to the chaos of pre-Westphalian days.

Some people panic over this rising mayhem, thinking that it forebodes the collapse of world order, but fear not. It's natural. This change is a resetting back to an old normal. Most of history is disorder. The past four centuries of a rules-based order governed by states is anomalous. Even then, it was hardly bloodless; World War I and World War II were the most devastating conflicts in history, judging by body count and urban destruction. Now we are returning to the status quo ante of disorder and what came before 1648. The world will not collapse in anarchy but smolder in perpetual conflict, as it has for millennia. We will be all right, if we know how to handle ourselves. The first step is realizing that no one fights conventionally anymore.

WAR HAS MOVED ON

"We have met the enemy and he is us." This play on Commodore Oliver Hazard Perry's famous words "We have met the enemy and they are ours" speaks volumes about how the strong lose to the weak today. Countries like the United States spend trillions of dollars preparing to fight the only type of war they will not face in the future—conventional war—leaving their citizens dangerously vulnerable.

As a young US Army cadet, I remember being taught how the Soviets would invade Europe. We spent hours studying a terrain map of the Fulda Gap, a spit of lowland on the border between East and West Germany, where NATO expected echelons of Soviet T-72 tanks to gush into free Europe. What made this odd was that the Soviet Union no longer existed.

"Sir?" I asked. "Why are we learning how the Soviets will invade Europe?" My mistake.

"Drop and give me fifty pushups!" barked the instructor.

"Forty-eight, forty-nine, fifty," I wheezed, knocking out my punishment in front of the class.

"Because future enemies will use Soviet tactics," he said, "and fight us the same way."

I doubt that, I thought but didn't say anything. Fifty more pushups were not going to improve my military education.

A year later, I was chuting up at Fort Bragg's green ramp. Lines of 82nd Airborne paratroopers were filing onto C-141 cargo planes. We were about to embark on a "mass attack" training exercise. Our mission was to parachute into hostile territory and seize an enemy stronghold, like D-day.

"Hell no!" someone yelled. A roar of paratroopers followed down the line.

"What's going on?" I asked my platoon sergeant.

"Someone got word from a buddy in the rangers. Said a bunch of rangers and Delta just got whacked in Somalia."

"*Somalia?*" I asked incredulously. America had the best military in the world. How did ragtag Somalis defeat our best-trained, most high-tech units: the 75th Ranger Regiment and Delta Force?

"Yeah. Somalia," said my platoon sergeant, spitting dip for emphasis. "Swarmed them and dragged their bodies through the streets. Fuckin' believe that shit?"

So much for Soviet tactics, I thought, as we loaded the aircraft for our D-day drill.

This was the famous "Black Hawk Down" incident of 1993.

Washington, DC, was so fixated on fighting the last war that it was blindsided by a stunning defeat. Tellingly, this defeat did not stop us from training against make-believe Soviet foes throughout the 1990s, when we should have been developing counterterrorism tactics for the 2000s.

What should we be doing for the 2020s? Beyond?

There is a saying: "Generals always fight the last war." This truism happens to be true. When it comes to seeing the future of war, nations turn to the past. Or rather, to past successes. We like to study victories that makes us feel good and ignore the unpleasant lessons of failure. This is how we get sucker punched by the future, usually at a heavy cost. On the eve of World War I, militaries were practicing Napoleonic horse drills, leaving them unprepared for the slaughter of the trenches. Afterward, the victorious Allies remained fixated on static trench warfare, then were blindsided by the blitzkrieg of World War II. Today, the US military prepares for the "Big War" against China or Russia, assuming it will be a conventional fight like World War II. Modeling the future on past glories ensures failure.

Nothing is more unconventional today than conventional war. Multiple studies confirm this trend: the number of unconventional wars has risen sharply since 1945, whereas conventional interstate conflicts are nearly extinct (see chart). In 2015, there were fifty armed conflicts in the world, but only one was a conventional war. Yet violence has not waned. Social science research shows that armed conflict has increased since the Cold War, and the number of conflict deaths in 2015 surpassed any in the post–Cold War period. As a retired US Marine general told Congress

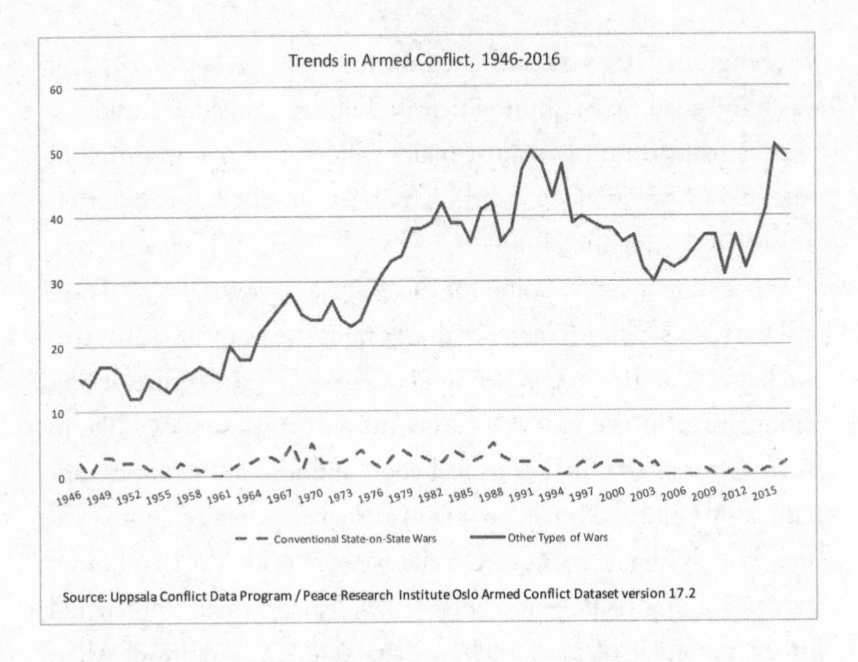

in 1999, "The days of armed conflict between nation-states are ending."

War has already moved on. Conventional war reigned uncontested from Napoleon to Hiroshima, lasting a mere 150 years. Despite the bitter lessons of Vietnam, Somalia, the Balkans, Iraq, Afghanistan, and Syria, the United States still maintains a military designed to thwart enemies who fight conventionally. America fights in the past, while its enemies wage war in the present. No wonder Afghanistan is the longest war in US history.

Conventional warfare is dead. Worse, it's killing us. This war orthodoxy squanders thousands of lives, trillions of dollars, and international standing as the United States struggles against weaker foes who do not fight like it's 1945. America is not alone, and other Western militaries suffer the same problem. The solution begins with the armed forces. We must retool the military to fight postconventional wars.

TRANSFORM THE MILITARY

There's an old joke: "How many nuclear submarines does it take to kill terrorists?"

"Not enough!" the military answers.

This is how strategic traditionalists think. When confronted with new threats, don't adapt—ignore! Perhaps this is why the US Congress appropriated funds in 2017 to buy thirteen more subs.

We shape our tools, and then they shape us. NATO is full of conventional militaries tooled to kill Soviets in a war that never happened decades ago. The Apache helicopter and the M-1 Abrams tank were designed to destroy Soviet T-72 tanks; the F-15 and F-18 fighter jets were created to shoot down MiGs; and carrier strike groups were organized to sink the USSR's navy. None of these weapons do much good against modern threats, yet we still buy them. Worse, we adapt our tactics to fit them, rather than the other way around. This is insane, but we do it because it's what we know.

First, we should stop buying conventional war weapons. In 2016, the United States spent $177.5 billion on the acquisition of such weapons—three times the United Kingdom's entire defense budget. About half of this money went to weapons that cost billions to develop and procure. The F-35 is one example. These weapons do not defeat threats like terrorists, nor do they deter countries like Russia or China. Russia took Crimea already, and China is conquering the South China Sea. These legacy weapons are low bang for the buck.

Meanwhile, weapons that succeed in modern wars are neglected. Special operations forces are an example. They are overdeployed, everywhere, all the time, because they are effective. What's their secret? They are nontraditional fighters, which is what makes them "special." Yet they receive only a tiny 1.6 percent of the Pentagon's half-trillion-dollar budget. Buying an aircraft carrier—a single ship—costs more than all US special operations forces combined. The navy wants to buy two more aircraft carriers, when in reality it needs to expand its special operations forces.

We need to rebalance the US military force as a whole, to contain fewer obsolete conventional weapons and more special operations forces and other tools that succeed in modern and future war. This investment will also save taxpayers oodles of money, because old-fashioned weapons systems are obscenely expensive. Even tripling the budget of special operations forces, which costs about $10 billion annually, is just a rounding error for traditional war's big-ticket items. Money saved should be reinvested elsewhere, and Americans would probably enjoy a hefty tax break as well.

However, special operations forces cannot be mass-produced or hastily created during an emergency. Lowering the bar for SEALs and other elite warriors is not the answer. Fortunately, it does not have to be. The whole military should become more like special operations forces, and future training should reflect this. Tactics, techniques, and procedures common in special operations forces should become more mainstream, especially among ground forces. Other unconventional capabilities should be expanded and improved, like psychological operations and civil affairs, which can mobilize people on the ground and build proxy militias for US interests.

Other bold steps should be taken, such as fundamentally realigning the military's force structure. Currently, the military keeps most of its conventional firepower, like tank divisions, on active duty. These full-time units train at military bases around the world and stand ready to deploy at any time. By contrast, support units remain in the reserves and mobilize when there is a national emergency. They perform tasks like intelligence, engineering, medical support, logistics, and a myriad of highly technical skills needed to sustain the warfighter. Reservists are part-time soldiers who are otherwise civilians, with peacetime careers outside the military. As a result, their military skills get rusty.

This arrangement of active and reserve components is exactly backward. Half of the US Army is in the reserves—the wrong half. Conventional warfighters should be sent into the reserves, and support functions moved to active duty. This realignment reflects reality. When a tank unit deploys, it will rarely perform tasks related to its core mission: destroying enemy tanks in a large land battle. Instead, it will do other things, like train foreign troops or engage in disaster relief. Meanwhile, support units are overdeployed, as they provide intelligence, medical support, logistics, and other tasks increasingly vital to modern warfare. A Department of Defense study during the Iraq War found that the reserves could not keep up with operational requirements. The few support units on active duty are overtaxed, conducting never-ending missions around the world. Support units are the backbone of modern military operations and will be more important than tank treads in future wars.

Change happens from the bottom up. Altering what units are in the reserves as opposed to on active duty sends a message to all ranks: support brigades eclipse conventional firepower for coming conflicts. Fair or not, reservists are deemed second-rate soldiers by careerists. Moving the famed 1st Infantry Division into the Kansas National Guard will send shockwaves throughout the system. So will standing up a C-17 cargo plane unit instead of an F-35 squadron. Not only will such moves prove more useful in future fights, but they will shape our future leaders. Ambitious young officers choose fields they think will get them promoted, and Pattons are not discovered in the reserves.

Revolution also occurs top-down, and this requires changing the generals' guard. The four-stars who run the US military are always conventional warriors, a Cold War legacy. In the army, they're the infantry and tank drivers. It's submariners and fighter pilots in the navy. For the air force, it's the fighter jocks. For the marines, it's a marine. This exclusive selection of the top-ranking general officers from among the conventional war branches makes no sense. If we are to face the future, the top billets must go to those who know how to fight in this new age. Generals drawn from special operation forces, intelligence, and information operations could break the old paradigm. Thinking trickles down the ranks, as with everything else in the military, and the whole institution could right itself in a generation if its leadership prepared for postconventional wars.

Special operations forces, or "SOF," also need rebalancing. Ever since the 9/11 attacks, elite units have honed the way of the knife but have let other skills dull. Those who hunt and kill terrorists are nicknamed "Black SOF," owing to covert actions like the raid on Osama bin Laden's compound. However, we need more "White SOF." They train allies' security forces to kill bad guys, so we don't have to, and this makes them a force multiplier. The US Army Special Forces, also known as the Green Berets, specialize in this mission, and we need more of them. Otherwise we will be forever chasing terrorists, and the numbers are against us.

New capabilities are needed, too, and countries should invest in warrior-diplomats in the mold of T. E. Lawrence. Green Berets do some of this, as do others. For example, the US military trains foreign area officers. It also has the Afghanistan-Pakistan Hands Program, which specializes in South Asia. This model is instructive. In it, individuals are immersed in a region for years, learning the local language, culture, and politics, as well as how and why the locals fight. Many live among the locals and even fight by their side, as Lawrence did. Not only can such people shape events on the ground for US interests, but their local expertise perhaps could have forestalled the worst of Iraq and Afghanistan, such as the de-Ba'athification of the Iraqi army that fueled the insurgency. Sadly, these warrior-diplomats are overlooked by top brass because their careers are too unconventional. We need more warrior-diplomats, not just on the ground in strategic locations, but also in leadership positions inside the Pentagon.

Lastly, not every warrior wears a military uniform, and money cut from huge conventional weapon systems should be reallocated to expeditionary civilian agencies. If firepower were all it took to win modern wars, then America should have conquered Iraq and Afghanistan easily. But war no longer functions this way. Other instruments of national power must be cultivated, such as information dominance, pinpoint sanctions that financially strangle enemy elites, strategic messaging to win the battle of the narrative, public diplomacy that speaks directly to populations, force that provides plausible deniability, and bribes to change adversaries' minds so that we don't have to shoot them. This is the work of the intelligence community, the Treasury and State Departments, and the US Agency for International Development (USAID). Much of our future arsenal already exists in the hands of civilians, not among troops, and strategic mavericks understand this dynamic. As General James N. Mattis testified before Congress: "If you don't fund the State Department fully, then I need to buy more ammunition, ultimately."

Preparing for conventional war is unicorn hunting. Worse, it's causing the West to lose. Typically, it takes a national near-death experience for a country to revolutionize the way it fights. In this way, the West is a victim of its own success, but others are now lapping it. Soon it will be left behind.

We should not wait for an existential crisis to force a change. We should begin adapting now.

RULE 2: TECHNOLOGY WILL NOT SAVE US

The F-35 fighter jet is awesome.

It looks like a spaceship, has 43,000 pounds of thrust, a top speed of Mach 1.6, and can perform instantaneous high-g turns. One model of the F-35 can even take off and land vertically, like a helicopter. But this fighter is no flyweight; it can drop nine tons of smart bombs into a shoebox.

Enemies will never know what hit them, because the F-35 is invisible to radar. It is the stealthiest war machine ever made, able to see without being seen. Anyone taking on an F-35, said one pilot, would be like jumping into a boxing ring "to fight an invisible Muhammad Ali."

The F-35 also gives its pilot X-ray vision. The helmet visor allows the pilot to "see through" the aircraft, using remote sensors on the fuselage, for next-generation dogfights. It may also be the smartest weapon ever made, running on eight million lines of computer code, more than the space shuttle. When the F-35 flies, its silicon brain can see the entire battle space, automatically select targets for the pilot, and link to other friendlies—tanks, destroyers, drones, missiles, launchers—for cooperative targeting.

The F-35 is a technological marvel, one that drives fighter pilots gaga with *Top Gun* lust. Even better, the plane practically flies itself. There are few gauges, buttons, or knobs in the cockpit. "What you have in front of you is a big touchscreen display—it's an interface for the iPad generation," said one test pilot. Move over Maverick and Goose, the F-35 is the future. Or, as another pilot said, it is "a needed aircraft to get us to where we need to be for the future of warfare."

The F-35 will be worthless in future wars. In fact, it is obsolete right now. There has not been a strategic dogfight—a sky battle that helped decide a war—since the Korean War, so why are more fighter jets needed? America has been at war continuously since September 11, 2001, yet this expensive superweapon has flown zero combat missions. The measure of any weapon's value is its utility.

The F-35 is a monument to our faith in technology, as evidenced by how much has been invested in it so far. It is the most expensive weapon in history. The United States has sunk $1.5 *trillion* into this airplane—more than Russia's GDP. If this plane were a country, its GDP would rank eleventh in the world, ahead of Australia's and Saudi Arabia's. Buying one costs around $120 million, double the price of a Boeing 737-600 airliner. They are also expensive to fly. Each hour in the air costs $42,200, more than double that of the F-16 fighter jet or the salary of the mechanic fixing it. Is this cost worth it for a fighter jet that never goes to war? Of course not.

The irrationality of the F-35 goes beyond its price tag—the plane is redundant. It was devised as a flying Swiss army knife that could meet the needs of the air force, the navy, and the army. Instead, it proves the adage that a camel is a horse designed by committee. It is true that the F-35 can engage in dogfights, drop bombs, and spy—just not well. Older aircraft remain better than F-35s at all these tasks. For instance, the A-10 Thunderbolt, an aircraft introduced in 1977, is better at ground-support missions (something the F-35's team admits). Dedicated bombers can also can fly farther with larger payloads.

Astonishingly, the F-35 cannot dogfight, the crux of any fighter jet. According to test pilots, the F-35 is "substantially inferior" to the forty-year-old F-15 fighter jet in mock air battles. The F-35 could not turn or climb fast enough to hit an enemy plane or dodge enemy gunfire. Similarly, the F-35 struggled against a 1980s-vintage F-16. The older aircraft easily maneuvered behind the F-35 for a clear shot, even sneaking up on the stealth jet. Despite the F-35's vaunted abilities, it was blown out of the sky in multiple tests.

Those who live by technology die by it, too. Unsurprisingly, the F-35's eight million lines of code are buggy, as are the twenty-four million lines running the aircraft's maintenance and logistics software on the ground. Sometimes pilots have to press Ctrl+Alt+Delete while in flight to reboot the multimillion-dollar radar. The F-35's computer code, government auditors say, is "as complicated as anything on earth." And what can be coded can also be hacked, another vulnerability for the F-35.

Hacked or not, buggy computer code grounds aircraft. The military's director of operational test and evaluation found twenty-seven "Category I" errors with the F-35's computers. These errors may cause death, severe injury, major damage to the aircraft itself, or "critically restrict the combat readiness" of the military. Grounded aircraft have zero combat effectiveness. After $1.5 trillion spent, twenty-five years of development undergone, and no combat missions flown during two long wars, it is incredible that the military still wants to purchase 2,443 of these things. Even the Pentagon's chief weapons buyer admits that the F-35 is "acquisitions malpractice."

It is not just the F-35—it's everything. People want cool stuff, rather than weapons that work. This is technological utopianism, and it is part of the Western way of war. Aircraft carriers do not defeat threats like ISIS, yet the United States launched another one in 2017, which cost $13 billion. Its purpose, in the words of President Trump, who launched it, was to "demolish and destroy ISIS." Ironically, this ship cost more than the entire budget of US special operations forces, which *are* effective against ISIS. The US Navy is ordering nine more such ships, too, even though carrier warfare hit its climax in 1942, at Midway. Outdated strategic thinking costs us all.

The opportunity costs are astounding. Taxpayers are swindled for trillions on weapons that are irrelevant, like F-35s, leaving the country dangerously exposed. What are we neglecting to buy that we need for future wars? Some experts believe that America's debt may be its biggest enemy. "No nation in history has maintained its military power if it failed to keep its fiscal house in order," said James Mattis, the secretary of defense and retired four-star general.

Future wars will be low tech. Cheap drones can be purchased off the internet and modified in someone's garage. Rig them with primitive explosives, and voilà! Your own kamikaze air force. Several hundred can be controlled at once, swarming a target, or guided by a GPS system like a Tomahawk missile. Modern enemies weaponize the mundane: commercial airliners, roadside bombs, suicide vests and trucks. Nations spend billions of dollars trying to defeat these crude weapons, and fail. Low tech—so easy to obtain and so difficult to defeat—will form the future's weapons of choice.

Meanwhile, F-35s sit on runways around the globe for years, as the United States fights in Afghanistan, Iraq, Libya, Syria, and elsewhere. These fighter jets are obsolete because they were built to fight conventional wars in a postconventional age.

THE WAR ALGORITHM

Faith in technology, such as the F-35, is a form of self-deception. Today's conflicts already show us that superior technology and firepower do not win wars—as confirmed by the F-35's war record. Yet Washington is seduced by its sexiness, and military planning often descends into tech porn.

Not long ago, I was at a Washington, DC, think tank event on the future of war. Hill staffers, military officers, foreign policy gurus, and big defense contractors made for a standing-room-only audience. TV cameras lined the back wall, and the smell of cheap coffee filled the air. There was a new catchphrase going around town—"Third Offset Strategy"—and everyone was eager to learn more.

As I took my seat, I overheard two defense analysts next to me. "The problem," one said, "is that the military can never fully anticipate tomorrow's threats. However, it can future-proof itself through technology."

"That's why DoD needs more money," the other said, referring to the Department of Defense. "The military is woefully underfunded." Both nodded sagaciously.

Chatter gave way to silence as Robert O. Work, the deputy secretary of defense, took the podium. After pleasantries, he got to his point.

"It's become very clear to us that our military's long-comfortable technological edge—the United States has relied on a technological edge ever since, well even in World War II. We've relied upon it for so long, it's steadily eroding." He went on to explain how the United States' "competitors" from North Korea to Russia were "devising ways to counter our technological overmatch."

The solution? More technology. Work explained how superior technology can "offset" any conceivable threat, no matter who, what, when, where, and why. This is the Third Offset Strategy.

"Let me just quickly go through what I mean by an offset strategy," he continued. The First Offset was nuclear weapons in the 1950s; the Second Offset gave us precision-guided munitions in the 1980s; and now the Third Offset would bestow the promise of robotics and artificial intelligence. It was a revisionist's military history, but never mind. The crowd was eating it up.

There was nothing new in this strategy. It was a reboot of the old Revolution in Military Affairs theory from the 1990s that gave us the failed "Shock and Awe" campaign during the Iraq War in 2003. The Revolution in Military Affairs proved a spectacular failure, but people overlook inconvenient facts when drunk on confirmation bias. Olson's Rule for Washington: You don't have to be good, you just have to be plausible. People forget the rest.

Over the next few months, Work and other Pentagon officials proselytized the Third Offset Strategy and faith in technology.

"Brothers and sisters, my name is Bob Work, and I have sinned," he said to laughter at one event. "I am starting to believe very, very deeply that [technology] is also going to change the nature of war." Sensing the gravity of this statement, he added, "There's no greater sin in the profession."

Suggesting that new technology can change the immutable nature of war rather than just how it's fought is ignorant. That's like claiming a new kind of clock will change the nature of time. Yet many in military circles are starting to believe that artificial intelligence will redefine war because thinking would be faster than thought.

"Learning machines literally will operate at the speed of light," Work said. "So when you're operating against a cyberattack or an electronic attack or attacks against your space architecture or missiles that are screaming in at you at Mach 6, you [need] . . . a learning machine that helps you solve that problem right away."

The military would produce, Work promised, human-machine mind melds. Weapons of the future would be like centaurs, part man and part machine.

"The best example of this," he continued, "is the F-35 Joint Strike Fighter. We believe and we say it over and over, this fifth-gen fighter, [even though] it can't out-turn an F-16 or . . . go as fast, we are absolutely confident that F-35 will be a war winner . . . because it is using the machine to help the human make better decisions."

A colonel next to me whipped out his iPhone and googled the F-35. Pictures popped up showing the plane doing high-speed things, like flying upside down. Many pictures were just artists' sketches rather than photographs. One depicted the F-35 leading a squadron of drones into battle, flying seven hundred miles per hour, all controlled by the centaur mind of man and machine.

Such is the stuff of fantasy, not beta testing, but Work's shock and awe campaign was working.

"Cool," whispered the colonel. "This is so cool."

"I'm telling you right now," Work said, "ten years from now if the first person through a breach isn't a friggin' robot, shame on us."

The colonel was sharing his iPhone with his neighbors, who were equally impressed.

The presentation ended with a shout-out to the defense industry, whose representatives nearly stood up and whooped. The Third Offset Strategy bestows a multibillion-dollar shopping list to the military-industrial complex. To push matters along, the Department of Defense set aside $18 billion as seed money. It also took the unprecedented step of establishing its own venture capital fund in California's Silicon Valley, courtesy of US taxpayers. It need not be profitable, just supply the warfighter with gee-whiz technology. They call it the Defense Innovation Unit-Experimental, or DIUx, and its slick home page features an "ENTER A $100+ BILLION MARKET" button front and center.

Work ended with, "We will kick ass." People applauded like it was Broadway. Military visionaries have seen the future, and it is iCombat. A month after Bob Work left the Pentagon, he joined the board of directors of Raytheon, one of the biggest defense contractors in the world, with $22.3 billion in revenues. Just another brick in the wall of the military-industrial complex.

No idea is so wrong that it can't find someone to believe in it. In 2017, the Pentagon initiated Project Maven. Its mission is to develop "algorithmic warfare" and win the artificial intelligence arms race. Imagine a megacomputer so powerful it can predict wars before they occur with near-perfect accuracy, allowing preemptive strikes against would-be attackers, similar to Philip K. Dick's short story "Minority Report." In the story (and in the Tom Cruise movie that followed from it), future society possesses a machine that allows the police to see future crimes, allowing them to arrest the perpetrators before an offense occurs. Often the arrestees are bewildered, screaming their innocence as the police drag them to life lockup for a crime they have not committed, but supposedly will.

Project Maven operates by the same logic as that of "Minority Report." Call it the war algorithm, a case study in modern war ethics and the violation of privacy rights. It would take the guesswork out of the future by sucking in every email, camera feed, broadcast signal, data transmission—everything from everywhere—to know what the world is doing, with the omniscience of a god. The ancient Greeks had a word for this: hubris.

A year later, thousands of Google employees signed a letter protesting the company's involvement in Project Maven. "We believe that Google should not be in the business of war," the letter said, addressed to Sundar Pichai, the company's CEO. Resignations followed.

Sometimes the more obvious something is, the greater the likelihood is that it will be overlooked. Sexy technology does not win wars. Since World War II, high-tech militaries have been thwarted consistently by low-tech opponents. The humble roadside bomb still outsmarts America's smart weapons, and the lowly AK-47 is the world's true weapon of mass destruction, if measured by people killed.

INVEST IN PEOPLE, NOT MACHINES

We have invested in technology at the expense of our people. Here's but one example.

At one thirty in the morning, sailors aboard the USS *Fitzgerald* were jolted awake in their bunks. Some were thrown to the floor. A container ship three times their tonnage had slammed into the destroyer off the coast of Japan.

Cold water gushed into their living quarters, two decks below the waterline. One sailor was knocked out of his bunk by the deluge. Within a minute, the seawater was waist deep.

"Water on deck!" sailors shouted. "Get out! Get out!"

Mattresses, furniture, and an exercise bike floated down a corridor. The power cut out and the emergency lights turned on, but it was still dark.

"Clear the compartment!" someone yelled. "Check racks. Make sure everyone is out."

Sailors waded through the darkness and around pieces of mangled steel. Moments later, the water reached the ceiling. One sailor survived by breathing in a small air pocket and then swimming toward the starboard egress ladder.

As the water forced them up the stairs, two sailors chose to remain below, fishing out two more shipmates.

The last survivor was in the bathroom at the time of the collision. The force of water had thrown him to the floor, and the compartment had flooded in less than sixty seconds. Desperate, he scrambled across floating lockers toward the main berthing area. As the water rose, he was pinned between a locker and the ceiling, until he was subsumed.

Grabbing hold of a pipe, he yanked himself free and swam toward the only light he could see. But it was too far. His breath ran out and he involuntarily inhaled water, drowning. A hand grabbed him, pulling him up. He lay on the deck gasping, eyes bloodshot and his face red.

The lucky ones escaped Berthing Compartment 2, but some were trapped in the debris as the water went over their heads. Others were crushed outright when the container ship's underwater bulbous bow rammed through the destroyer's hull. The area flooded in less than a minute. Of the thirty-five sailors who were in Berthing 2 when the destroyer was struck, twenty-eight escaped the flooding, and seven died.

The ship's commanding officer was also trapped in his cabin, wreckage blocking the door. It took five sailors, a sledgehammer, a kettlebell, and their bodies to break down the door and rescue him.

The *Fitzgerald* settled into a 7-degree list to her starboard side. Sailors rushed to seal off the flooding and keep the ship afloat, as water gushed through the thirteen-by-seventeen-foot hole beneath the waterline. Large pipes that carried seawater to fight fires also ruptured, inundating other areas. The *Fitzgerald* was sinking.

"The crew of the *Fitzgerald* fought hard in the dark of night to save their ship," the navy's report later said. The destroyer limped back to Yokosuka naval base in Japan. Dockworkers must have gasped at the sight: the superstructure on the starboard side was mashed in, just under the bridge, where the larger freighter, the *ACX Crystal*, had struck the *Fitzgerald*. The steel-plated hull was bent inward, and a mess of pipes, wires, and debris stuck out a few feet above the waterline. Below the waterline was a hole in the hull.

Just inside were the bodies of the seven drowned sailors, trapped in the wreckage.

The navy relieved the *Fitzgerald*'s captain, executive officer, and senior enlisted leader of their positions. The charge: "loss of confidence" in their ability to lead. Although none of them was present at the control stations when the collision occurred, they were held accountable for the failures of the crew.

The *Fitzgerald* accident is part of a pattern. Just nine weeks later, the USS *John S. McCain* collided with an oil tanker east of Singapore, also at night. Ten sailors died in the flooded quarters below deck. Earlier that same year, the USS *Lake Champlain* had collided with a seventy-foot fishing vessel. And a few weeks before that, the USS *Antietam* had run aground off the coast of Japan, dumping 1,100 gallons of hydraulic oil into Tokyo Bay.

Ships of the line do not do this.

In response, the navy took the rare step of issuing a worldwide twenty-four-hour "operational pause," or "safety stand-down," for the fleet. Crews reviewed teamwork, safety protocols, seamanship, and other "fundamentals" aboard 277 vessels. The navy also ordered a sixty-day top-to-bottom safety review of fleet operations, training, and certification. Lastly, it fired the three-star admiral who commanded Seventh Fleet in the Pacific.

None of these steps solves the root problem. The root problem is that of untrained personnel. These are the most expensive and technologically advanced ships afloat. How did they collide with other ships in the middle of an ocean? Technology did not save them.

The solution is investing in people, not platforms. Gray matter is more important than silicon, and focusing on people rather than hardware should be the highest defense priority. This is not what the defense community is calling for today, as evidenced by the Revolution in Military Affairs, the Third Offset Strategy, and whatever tech craze comes next.

Overreliance on technology is an easy crutch that is dumbing down the force. The navy demands that its sailors master advanced systems at the expense of basic seamanship, such as the ancient art of celestial navigation. When you drive a car today, it is easy to use GPS navigation automatically. Should it fail, many people would be lost. It is the same for high-tech militaries. Fundamental skills suffer in the pursuit of high-tech goals, something future enemies will exploit. For example, knocking out GPS will offset advanced militaries. Future enemies will employ strategies of technological denial and attack.

The problem starts at the top. According to one retired navy captain, the admiralty believes "there is a technical solution, and we are looking to industry to provide a solution." Not only does this blind faith in the industry produce duds like the F-35, but it also produces a new generation of officers who lack knowledge of the fundamentals of their profession, such as seamanship. "A radar can tell you something is out there, but it can't tell you if it's turning," said the captain. "Only your eyes can tell you that. You have to put your eyes on the iron."

Once upon a time, navy surface warfare officers, or SWOs (rhymes with "goes"), spent their first six months going to school in Newport, Rhode Island, learning basic seamanship, leadership, and how to pilot a ship. Now they get CDs. Starting in 2003, each young officer instead has been issued a box set of twenty-one CDs for computer-based training, jokingly called "SWO in a

Box." If the young ensigns have questions, there is no instructor to ask. They are expected to master seamanship by laptop in between other duties while serving aboard a ship. Navy life is brutal, sometimes demanding workweeks of more than one hundred hours. No wonder destroyers collide at sea.

One young navy lieutenant learned of his sloppy seamanship the hard way during a two-year tour as an exchange officer aboard a British destroyer. "Frankly," he said, "I was embarrassed at my lack of maritime knowledge and skills for the first few months of my exchange. My first 90-minute-long written [maritime] Rules of the Road exam was a disgrace. I was accustomed to the U.S. Navy's 50-question multiple-choice exams." The Brits sent him to remedial training, which he welcomed. Royal Navy officers undergo a rigorous education in seamanship and other fundamentals, both on shore and at sea. They must also be certified by the International Maritime Organization's Standards of Training, Certification, and Watchkeeping. This robust training is a far cry from twenty-one CDs. The race for whiz-bang technology at the expense of proper training costs lives in peacetime. What will happen in war?

The navy finally learned its lesson, and it now requires some classroom time for new officers. But the damage is done. A generation of officers trained by CDs are commanding ships and mentoring the next generation. Worse, the tech-centric mentality still exists, as evidenced by the navy's embrace of all things technological and the Third Offset Strategy. Such thinking will lead to peril in the South China Sea.

Intelligent humans will always find a way to outfox smart weapons. Despite what technophiles claim, future wars will not be fought by robots. Human beings will matter more, not just morally but operationally, too, because technology is no longer decisive in modern war. The lessons of Iraq and Afghanistan scream this fact, as luddites bested the West's overwhelming military and technological superiority. This is true on the waves, too, as China shows. If anything, technology is dangerous—to us. It dumbs down the military, bankrupts the treasury, and leaves us critically unprepared for future conflicts, which will not be conventional wars.

Machines will not save us, but human beings can, and that's where we should invest. It's probably cheaper, too, by several decimal places. Again, gray matter is superior to silicon. Supercomputers who beat grandmasters at chess do not predict the future of war, only chess. War is infinitely more complex than a game. Perhaps when a supercomputer runs for president or prime minister, we should be concerned. Until then, it's hype. Those with a stake in the military-industrial complex will cry foul while wrapping themselves in the flag, but no one should fall for their con.

Technology is no longer decisive in warfare. The quicker we abandon our addiction to it, the sooner we will win wars again. It's strategically distracting. This does not suggest we forsake sophisticated gear, but we should stop worshipping it. Gizmos can shape our everyday lives, but not victory. War is armed politics, and seeking a technical solution to a political problem is folly. Ultimately, brainpower is superior to firepower, and we should invest in people, not platforms.

RULE 3: THERE IS NO SUCH THING AS WAR OR PEACE—BOTH COEXIST, ALWAYS

Somewhere in the South China Sea, the navy destroyer USS *Stethem* (pronounced "STEE-dem") prowls the waves. A few miles ahead of it lies a tiny island that is little more than sand and seagull dung. Mainland China is 386 miles to the north, monitoring the destroyer's every move.

"Come left, steer course zero six zero," orders the conning officer.

"Come left, steer course zero six zero, aye," repeats the helmsman, turning the ship's wheel.

The Arleigh Burke–class destroyer is the workhorse warship of the US Navy. Capable of operating independently or as part of a strike group, it can see everything below, on, and above the water's surface, including objects in space—and it can kill anything it sees. There's not a single 90-degree angle on the destroyer's superstructure, making it stealthy and hard to observe with radar. Propelled by a whopping 100,000-shaft horsepower into two screws, the 500-foot ship can achieve speeds over 30 knots, slicing through the waves with its raked bow, and it can stop from that speed within its own length. It is one of the most powerful warships in the world, and the United States has sixty-six of them.

The bridge is tenser than normal. But this is no routine mission. Down below, two large screens loom above an array of monitors in the combat information center. The screens track the ship's heading toward the island.

The captain, or "CO," is worried, although he would never reveal this to the crew. Their mission is to follow a track within twelve nautical miles of the island, defying Beijing's sham claim to it. China has said it will fire on any vessel that violates its sovereign territory, and it is the destroyer's mission to do so. It could mean war. In a game of diplomatic chicken, Washington is sending Beijing a message: You can't steal islands, no matter how small.

They are close. The conning officer scans the island with binoculars. Only 1.2 square kilometers, it's a speck of sea dirt by ocean standards. Chinese fishermen once called it Snail Island, but now it's known as Triton Island, part of the Paracel Islands archipelago claimed by Taiwan, Vietnam, and—only recently—China.

All is quiet. Then the radios burst alive. China warns the guided missile destroyer that it is within twelve nautical miles of Triton Island and is trespassing on Chinese territory. The US ship is instructed to turn back immediately.

Displays in the combat information center light up and alarms sound as China deploys military vessels and fighter jets. The crew of the *Stethem* remain unfazed. The two big displays show Chinese jets closing rapidly, tracked by the destroyer's AEGIS weapons control system.

"TAO, Air, heads up. New tracks 04876 and 04877, bearing

zero-three-five at forty-five nautical miles inbound. Angels Three, I assess as TACAIR based on speed and altitude," says the sailor working the air warfare console.

The "TAO," or tactical action officer, rushes over to see for herself. She's in charge of the combat information center, answering directly to the captain. Instantly, she recognizes the planes based on their flight signature: Shenyang J-11s, modeled after the Soviet Su-27 fighter. They are lethal.

"Things are starting to get a little sporty," the CO whispers to the TAO, seeing the same thing.

"Air, TAO," she says. "Cover tracks 04876 and 04877 with birds." The sailor working the console quickly rotates a large trackball while pushing orange buttons to the left of the screen. Each Chinese jet is assigned one of the ship's surface-to-air missiles.

"TAO, let's issue a query to the inbound aircraft if they break thirty miles," says the CO.

"Aye," she says.

On deck, sailors scurry about, a brisk wind cutting across the deck. Above them a gigantic American flag flutters, port side of the mast. They all wear blue uniforms and baseball caps with USS STETHEM stitched in gold on the brim. One sailor holds on to his cap so it doesn't blow away. Rows of missile hatches line the deck. Stenciled on the side is DANGER. STAND CLEAR OF LAUNCHER DECK AREA.

"Surface, TAO. Interrogative track 04845," says a sailor manning the surface warfare console, betraying his alarm. A new threat.

A Chinese destroyer, thinks the CO with a groan. They have stringent rules of engagement: Sail into harm's way and don't blow it. They are not here to hunt but to send a message. He prefers to hunt.

"Hold what you've got," the CO reassures the combat information center, as he presses the mic on his headset. "Bridge, CO. Make sure your response is in accordance with international law."

"Bridge, aye," responds the officer on deck over the net.

What are you up to? the CO thinks as he studies the tactical displays, undistracted by the various alarms. He has only a few minutes to discern China's intentions before the USS *Stethem* is in range of its missiles. The lives of his crew depend on it. If he doesn't defend his ship, they could be blown out of the water. But if he attacks first, it could spark a war between the United States and China.

The USS *Bedford*, a sister ship, had recently encountered a similar situation. He learned a lot over beers one night with its CO back at Yokosuka, Japan, their home port. What he heard had disturbed him. The Chinese liked to play for the edge.

"TAO, Air. Negative response to query. Track 04876 and company now at thirty-five nautical miles inbound, descending to angles 2.5. I intend to issue a warning at twenty-five." The Chinese jets were ignoring their warnings.

This could get more than sporty, the CO thinks. He glances at the electronic warfare watch station, prepared for signs of an incoming attack.

"TAO, EW. Negative ES from tracks 04876 and 04877," the technician says, reporting no indications that the aircraft are prepared to launch a cruise missile.

The jets are closing in.

"Hold what you've got," the CO repeats.

"TAO, Air. Track 04876 and company, negative response to

warning, continuing inbound. Distance now twenty miles, looks like they intend to mark on top. Continuing to warn."

Eighteen miles. Seventeen, sixteen, fifteen. The destroyer was in easy range of the jets' antiship missiles.

"TAO, Air. Track 04867 and company level Angels Two, fifteen miles inbound overhead in four minutes. I intend to illuminate at ten miles," the operator says, meaning he will "light up" the jets with a beam of intense radar used for guiding surface-to-air missiles to their targets.

"Air, CO. Negate illumination. They may interpret your warning as hostile and fire on us in self-defense," says the CO. This was the scenario he dreaded.

"CO, TAO. Tactical aircraft are now at ten miles. We will have video on them shortly, recording is on."

"Very well, but don't get too focused on the aircraft. Think big picture. What's your surface picture? Is there a sub stalking us? Don't forget about the Chinese fishing militia." The jets could be a decoy, meant to distract them from the real attack.

"Combat, Bridge," a voice crackles over the net. "Two small fighters low on the horizon inbound, look like J-11s maybe. Their wings look clean, hard to tell, could be fuel tanks."

"TAO, Guns. Video has them, two J-11s, wings clean air-to-surface," says a sailor, meaning the jets have no missiles or bombs.

The Chinese fighter jets screech overhead. Sailors on deck cover their ears. The flyby was aggressive. Sending unarmed jets, their lack of weapons discernable only at the last minute, was playing the edge.

"Just another day in the office, huh?" says the CO, patting the TAO on the shoulder as he makes his way to the bridge.

NONWAR WARS

Cunning adversaries leverage the space between war and peace for devastating effect. Washington has a buzz phrase for this: the "Gray Zone." Others have a strategy. In Russia, experts call it "New Generation Warfare," and it conquered eastern Ukraine and Crimea. Israel has the "Campaign Between the Wars" to punch back in a tough neighborhood. China's version is called the "Three Warfares" strategy, and it is how its leaders plan to drive the United States from Asia.

The Three Warfares strategy succeeds because it's war disguised as peace. To the West, the South China Sea looks paradoxically like a "nonwar war," and that's how China wants it. Yet it is proving more effective than a fleet of aircraft carriers. It wins by strategic jujitsu, using the West's paradigm of conventional war against it. Conventional warriors view war like pregnancy: either you are or aren't. War versus peace is enshrined in the Laws of Armed Conflict, the writings of Clausewitz, and conventional war theory.

For the West, war occurs when peace fails. Take World War II and the US Civil War. Japan broke the peace when it attacked Pearl Harbor, and the Confederacy declared war on the Union when it shelled Fort Sumter. Then war is declared and battle the great decider. In war, ethical standards lapse, and atrocities are committed. Conventional warriors justify these things as "collateral damage." War ends at the peace table, whether it is aboard the battleship USS *Missouri* or at Appomattox Court House. You behave differently in war versus peace, making the distinction between them crucial. China wins because it exploits the West's false belief in this dichotomy. Beijing knows Washington has a light bulb vision of war: it's either on or off. The trick is to keep the US war switch flipped to "off" so the superpower remains docile and at "peace."

China can get away with almost anything in the space between war and peace, and it does. How does it do it? Two ways. First, the Chinese play a dangerous game of brinksmanship, going right up to the edge of war—or what the West thinks is war—and then stopping. They escalate to deescalate and then keep what they capture (or create). This strategy works because they know America won't flip the war switch to "on" and risk a nuclear exchange over "a few scattered rocks in the Pacific," as one CIA director put it. If China does this enough times, it will eventually own the South China Sea.

Second, China masks its conquest using nonmilitary tools, so its actions don't look like war to conventional warriors. Strategic disguise is an ancient Chinese tradition, dating back to Sun Tzu and the Thirty-Six Stratagems. And the Chinese still practice it. In 1999, two Chinese colonels wrote the book *Unrestricted Warfare*, in which they outlined how China, with its inferior military, could best the West using psychological warfare, economic warfare, "lawfare," terrorism, cyberwar, and the media. The Three Warfares strategy, which builds on this idea, was then officially recognized by China's Central Military Commission and Communist Party in 2003.

Worried, the Pentagon commissioned a report on China's new way of war. The results were astonishing, at least to the conventional war mind. The Three Warfares strategy is designed to blunt US force projection through "war by other means" (bizarre Pentagon jargon for wars that don't prioritize bombs). However, China's strategic logic is compelling. It recognizes that conventional warfare is obsolete, and that the best weapon is not the military. According to the report, "In the modern information age, nuclear weapons have proven essentially unusable and kinetic force . . . [has] too often brought problematic outcomes and 'un-won' wars." China drew this conclusion in 2003, well before observing the US face-plant in Iraq and Afghanistan. For Beijing, military force wins in "ever decreasing scenarios."

Instead of battlefield victory, the Three Warfares strategy achieves victory by sapping the enemy's will to fight before the fighting even begins. It's pure Sun Tzu, and it does this using psychological warfare, propaganda, and legal tools of war. Psychological warfare targets the enemy's decision-making calculus, causing him to doubt himself and make big blunders. Its toolkit includes strategic deception, diplomatic pressure, rumor, false narratives, and harassment. When fighting democracies, it tries to stir up antiwar sentiment in members of the enemy's population, encouraging them to elect new leaders who see things Beijing's way. Psychological warfare is about making enemies second-guess themselves and depleting their will to fight.

When China wants to compel the United States to do something, it doesn't turn to force. That's junior varsity stuff. Instead, it messes with America's pocketbook and turns to psychological warfare, which is far scarier than a stealth jet. China threatens the sale of US debt, pressures US businesses invested in China's market, uses boycotts, restricts critical imports and exports, and employs predatory practices. This shrewdly pits the US business community against its government. In 2017, the Chamber of Commerce urged the White House to "use every arrow" in its quiver to ensure a level commercial playing field in China but added the caveat that "some of these points of leverage could be counterproductive to us." Translation: "Don't screw us . . . or else." The Chamber of Commerce spent over $82 million lobbying politicians in 2017, so when it speaks, decision makers listen. By setting Wall Street against K Street, Beijing undermines the United States' will to fight.

China's media warfare is even more pernicious. Through it, the country seeks to manipulate Western public opinion for Chinese advantage. Its weapons of choice are film, television, books, the internet, and the Xinhua News Agency, the mouthpiece of the Communist Party of China. The tip of China's media spear is the China Central Television network (CCTV), which has a major facility in Washington. Like Moscow's RT news, it is state-controlled media disguised as CNN or Fox, reaching forty million Americans along with hundreds of millions more viewers around the world. Chinese president Xi Jinping urged the network to "tell China stories well," according to Xinhua. China has also bought much of Hollywood, making it impossible to cast China as a villain in movies—a brilliant strategic move for the international court of public opinion.

Controlling the narrative of a conflict is important for winning modern and future wars, especially when those wars are fought against democracies. People vote for leaders based on what they learn in the news. Whenever an incident breaks out between China and the Philippines over disputed reefs, CCTV is there first, spinning China's version of events, often before the Western media are even on to the story. Other news outlets pick up the lead, and soon it becomes accepted fact. The United States must play catch-up, debunking China's aggressive narrative before credibly advancing its own. In a similar vein, when tensions mount over the disputed Senkaku Islands, CCTV will quickly launch a strong offensive blaming "right-wing nationalists" in Japan for any incident or escalation. CCTV gives China the first-mover advantage in the media wars, and truth is always the first casualty.

As for China's legal warfare, or "lawfare," its goal is to bend—or rewrite—the rules of the international order in China's favor. This is not the rule of law, but rather its subversion. International law is the glue that binds the society of states, and it constitutes the "rules of the road" for world affairs. China wants new rules made in its own image.

Beijing uses lawfare to bolster its territorial claims. For example, it legally proclaimed the village of Sansha on the disputed Paracel Islands to be Chinese, even though Vietnam, the Philippines, and other countries had claimed the islands much earlier. China runs off foreign ships within its self-declared two-hundred-mile exclusive economic zone, even though the UN Law of the Sea treaty forbids this. China justifies its actions using the law, but the justifications consist of legal fictions packed with outlandish arguments. The bigger the falsification, the more China defends its legality—over and over again—believing that with repetition, people will finally accept it. Hitler had a name for this technique. He called it the "Big Lie."

Lawfare is best used before the outbreak of physical hostilities, according to the Chinese military. Done well, it raises doubts among many in the enemy camp about the legality and legitimacy of military action, undermining their will to fight. Examples of lawfare include changing international law and creating a cadre of legal experts around the world sympathetic to China's cause. Other stratagems are more subversive. The success of US military operations abroad often hinges on access to foreign bases. China could hamstring US military deployments by filing legal motions in American courts aimed at delaying any intervention. Simultaneously, it could take parallel legal actions against US allies in the region, such as Australia, Singapore, and the Philippines. This would disrupt US force projection, resulting in a strategic win for China.

China also applies lawfare to outer space. Its legal scholars claim the country's terrestrial borders extend indefinitely upward and into the cosmos, making it China's sovereign territory. Perhaps China will launch a space police force to catch all those undocumented satellites from entering its country illegally. From China's perspective, lawfare is a form of combat against the international rules-based order, an arena in which it believes America makes the rules and others obey the orders.

The United States responds to China the only way it knows how, by building more conventional arms. Superweapons like Ford-class aircraft carriers, F-35s, high-tech drones, Zumwalt-class stealth destroyers, and rail guns are all meant to deter Chinese aggression, but they don't. Nor does the Third Offset strategy. Frustrated, the US military sits on the sidelines, waiting for the start pistol of war. It knows it would cream the puny Chinese military once the Big War was declared. To boost morale, generals assign reading lists that include fictional accounts of the Big War between the United States and China, in which America wins Tom Clancy style. US military leaders even call this fantasy the future of war. Meanwhile, China swallows up more islands, and allies question

US leadership in Asia. American strategists scratch their heads, wondering why tricks that worked in the Cold War now fail.

This is not just a US problem. The line between war and peace has become so blurred that all nations mired in the old rules of war are baffled, and they now rush to call everything an "act of war." The Iranian-backed Houthis in Yemen fired a missile at Saudi Arabia's Riyadh airport. "We see this as an act of war," the Saudi foreign minister, Adel al-Jubeir, told CNN. "Iran cannot lob missiles at Saudi cities and towns and expect us not to take steps." Trump's tweet-bombs are also an act of war, according to the North Koreans. Their foreign minister told reporters, "Since the United States declared war on our country, we will have every right to take countermeasures, including the right to shoot down United States bombers." The tweet in question called Kim Jong-un, the leader of North Korea, "Little Rocket Man." The United States also lobs the *W* word. Senator John McCain and other senior lawmakers have labeled Russia's interference in the 2016 presidential election "an act of war." The US ambassador to the United Nations, Nikki Haley, agreed, declaring that "when a country can come interfere in another country's elections, that is warfare." The striking thing about all these "acts of war" is that none have led to war.

EXPLODING HEADS

"Is this war? Is this peace?" asks a US naval officer, head cradled in his hands. "If it's war, I know what to do. If peace, that's something else. But it's neither. Or both. What are we supposed to do?"

Heads nod around the seminar room. They are my students at the National Defense University in Washington, DC: senior military officers, civilians from other places in the government, and a few foreign military officers.

"It's a diplomatic situation that only looks like war," says an army colonel. "It's a failure of diplomacy."

"Not so fast, buddy," replies a diplomat. "The Department of Defense always blames its problems on the State Department."

"Isn't it diplomats who start wars and soldiers who have to finish them? It's harder to end a war than start one. I know because I've lost guys. How many have you lost?"

"You know we're on the same team, right?"

The colonel rolls his eyes. The diplomat shrugs it off. The Departments of Defense and State have been having this same conversation since 2004, when everyone realized there was no postconflict plan for Iraq, and things turned FUBAR—a military acronym that means "fucked-up beyond all recognition." Finger-pointing is the bureaucratic way of war.

"What does winning look like in the South China Sea?" I ask, trying to keep the discussion on course. The group ponders the question for a moment and then everyone speaks at once.

"We obviously need more aircraft carriers, subs, and F-35s."

"We already have the superior military, yet here we are, having this conversation."

"Better trade. We become so economically interdependent that it would be irrational to go to war."

"Because that worked so well in 1914," someone adds sardonically.

"It's not war, so it's not a question of winning."

"It may not be war, but we can still lose."

"It's the Kobayashi Maru," someone jokes, making an obscure Star Trek reference for a no-win situation.

"Yeah, but Captain Kirk won by changing the rules," says another, laughing. "We need new rules for war."

"Winning means upholding a rules-based international order."

"And reassuring allies in the region," adds the Filipino colonel. "The US will lose its leadership if it does not back allies."

People nod.

"But how far are we willing to go?" asks a marine. "Nuclear war?"

"It will never come to that."

"Why not?"

The room falls silent once more.

"Our job is not to fix the world," says a SEAL commander. "It's only to keep it from blowing up."

"What do you mean?"

"It's the Superman strategy. When an airliner is about to crash, Superman swoops in and saves it. He gets everyone safely to the ground but doesn't stick around to rebuild the plane, like we tried to do in Iraq and Afghanistan. He goes off and saves other crashing planes."

"I second that," says the air force pilot with a smirk.

"So how does that apply to the South China Sea?" I ask.

"There's only so much the US can and should do," the SEAL says.

Some students nod while others shake their heads.

"China is a rising superpower and we're the reigning one. War is inevitable."

"No, it's not."

This debate continues for a while, until it fizzles. Where else can it go? The conversion turns back to the South China Sea, and everyone looks at the floor, stumped.

"The problem is that victory in the South China Sea is like finding a needle in a haystack when every problem looks like hay."

"The Chinese seem to be doing fine in the haystack," says the navy officer.

"The real problem is the US is playing chess while China is playing go," says the diplomat, referring to the ancient Chinese game. Like chess, go is easy to learn but hard to master. But go is more complex and requires greater patience, giving new meaning to the phrase "playing the long game."

"I agree," says the army colonel, fist-bumping the diplomat.

Versions of this conversation have echoed across Washington for over a decade. Not much has changed, other than more Chinese islands in the South China Sea. Whatever the United States is doing, it's not working.

China is fighting in the twenty-first century, whereas America is stuck in the twentieth. The Three Warfares strategy conquers with a creeping expansionism designed to remain below the United States' threshold of "war," knowing that America will remain inert if at "peace." By viewing the South China Sea as a traditional military battlefront, the United States falls into China's strategic trap. Nonwar war is paradoxical to conventional war thinkers, and seeing them contemplate it is like watching a dog trying to pick up a basketball.

In modern and future wars, there is no war or peace—only war *and* peace. Those who grasp this will conquer, like China, and those who don't will speak Chinese. Expecting conflicts to formally start and stop is waiting for Godot. Bin Laden did not officially declare war before 9/11, and ISIS did not send an ambassador to sue for peace. That way of warfare is dead.

Because war and peace coexist, conflicts hibernate and smolder. Occasionally they explode. This trend is already emerging, as evidenced by the increasing number of "neither war, nor peace" situations and "forever wars" around the world. This is durable disorder. Protracted wars are the norm in history: the Hundred Years' War, the Thirty Years' War, the Peloponnesian War, the Crusades, the Roman-Germanic Wars, China's Warring States period, the Arab-Byzantine wars, the Sunni-Shia wars, and just about everything else. Even world wars I and II are best seen as a single war lasting thirty years. War and peace have always been an illusion. Can true peace really happen?

GRAND STRATEGY

The problem is that the United States has tied its end state to maintaining global peace and prosperity but has failed to conjure a meaningful grand strategy to achieve this. There is also another problem.

"Grand strategy is the biggest crock of shit."

"Say again?" I ask, nearly dropping my sandwich.

"Grand strategy is a myth. It doesn't exist," says Dan, a colleague at the RAND Corporation, a Department of Defense–sponsored think tank.

"Those are fighting words," I say, half-seriously.

"Bring it on," he replies, slurping his sweet tea with a smirk.

We are sitting on a park bench in the middle of the Pentagon courtyard eating lunch. It's the only park in America that has no squirrels. At the center is a small building that has fueled conspiracy theorists for generations. In reality, it's a café that was nicknamed "Ground Zero" during the Cold War, and it's where we picked up our lunch.

Dan is no strategic lightweight. He's a combat veteran with an Ivy League pedigree and a doctorate. More important, he has a fine mind. When he speaks, I listen, even though I don't always agree with him. This was one of those times.

Grand strategy is imperative, albeit hard to achieve. Done well, it guides a nation and synchronizes its actions in foreign affairs, but in recent years the idea has produced a lot of grand skepticism. This is dangerous. Grand strategy is like a rudder: without one, a country goes strategically adrift in the tumultuous sea of international relations. Most inside the capital Beltway believe the United States has been bobbing rudderless since the Cold War, and the same could be said of many countries.

Dan's cynicism is understandable because the concept of grand strategy has been corrupted for years. Abuse comes in four varieties. The first is "grand fluff": superficial babble that masks the absence of thought. Western countries commonly produce grand strategy that reads like political applause lines rather than an effective plan. In the United States this is the National Security Strategy. It has become a Santa's wish list of things America wants—world peace, no poverty, democracy everywhere. But you can't hand this strategy to a general or civil servant and say: "confront climate change" or "champion aspirations for human dignity." It's fluff.

Another source of grand skepticism is the Washington memoir. This vehicle of self-aggrandizement is part tell-all, part smear campaign against rivals, and part rationalization of past gaffes. They commonly emanate from foreign-policy titans, some of whom have written more than one. Sometimes their authors justify past bloopers by claiming they were following a grand strategy invisible to the masses at the time, but these justifications are kabuki theater.

The third source of grand confusion is fuzzy scholarship. Historians are apt to look into the past and identify an empire's grand strategy, even though the people of the time never would have recognized it as such. For example, ancient Rome never had a unifying grand plan the way we think of grand strategy now. But historians often attribute one to it because Rome was a millennial civilization, and wrongfully assume it must have possessed some unifying master plan. Political scientists are worse. One prominent scholar showcases the grand strategies of individual presidential administrations, missing the entire point of what makes a grand strategy "grand": it endures over decades and is not linked to a single leader—otherwise it's just petit strategy. Such a mix-up is a rookie mistake.

Mistaking bureaucracy for strategy is another source of grand confusion. Bureaucracy is a zero-sum game, except the sums don't add up. When the Pentagon gets its paws on the National Security Strategy, it breaks it down into constituent parts. Huge staffs perform an "objectives crosswalk" to derive a National Defense Strategy that derives a National Military Strategy that derives a Joint Operations Concepts (called the "Jopsy") for the "battle space." There are meetings for meetings, and colossal PowerPoint presentations written in nine-point type. When the giant staff finishes, three years later, the National Military Strategy is roadkill. Unrecognizable. Like a good joke overanalyzed, grand strategy loses its potency with too much processing. Grand strategy is a living idea that needs flexibility to breathe, and those who dissect it will kill it.

Despite this skepticism, grand strategy is real and urgently needed. Without it, the ship of state is rudderless in international relations. Certainly many Westerners feel we have been strategically adrift in recent years. The absence of competent grand strategy is why.

In its broadest sense, grand strategy is a policy that governs how a country behaves in international relations. States do not always make their grand strategies clear, write them down in one place, or even use the term to describe them. Sometimes a grand strategy is just a core set of beliefs about national security held by the foreign policy and military elite. ("Consensus" would be too strong a word.) These beliefs focus on what "security" means to the nation and how it should be achieved over the long term.

The purpose of grand strategy is to identify and protect national security interests. Not all interests are equal. Typically, they are differentiated among vital, important, and peripheral. Vital ones are existential; if we don't secure them, we die. Important interests can become vital ones if neglected, and peripheral interests are optional. In the US context, foreign policy debates among presidential candidates often focus on the racking and stacking of these interests. Democrats have one prioritization, Republicans another.

Good grand strategy has five characteristics. First, it is not restricted to war and recognizes that war and peace coexist. Second, it is dynamic and flexible, requiring a constant balancing of resources needed to ward off new threats. Grand strategy is not a checklist, but rather a jazz improvisation. Third, it harnesses *all* of a nation's instruments of power, not just the military ones. Fourth, it can be offensive or defensive. For example, the US policy of containment was largely defensive, while the Nazi idea of Lebensraum was offensive. Fifth—and most important—grand strategy endures over a long time, lasting decades or centuries. There is no such thing as a single administration's grand strategy. Successful grand strategies endure regardless of political party, individual leaders, or regime.

When I teach grand strategy at Georgetown University's School of Foreign Service, many young people have a hard time grasping it because they have never seen it in their lifetimes. Fair enough. The most recent example of a grand strategy in America occurred before they were born. During the Cold War, the US employed a grand strategy known as "containment." Its objective was to enclose communist expansion and to roll it back when possible. The foundational ideas emerged in George F. Kennan's "Long Telegram" and a real National Security Strategy called NSC-68. It lasted for fifty years, from 1950 to 1990, and although every administration had a different interpretation of it, its core strategic logic endured.

Containment reminds us what grand strategy entails. It had four components. First, it sought to maximize US influence and minimize the USSR's influence abroad. Second, it avoided direct confrontation with the USSR to avert nuclear war. Third, it tried to prevent a regional "domino effect" favoring the USSR. Fourth, it contained communist expansion through a variety of discrete strategies, such as nuclear deterrence through mutually assured destruction (or "MAD"), security cooperation efforts like NATO, coercive diplomacy, covert operations, proxy wars in places like Korea and Vietnam, the "rollback" of communist governments through regime change, and aiding democratic nations, also known as the Truman Doctrine. These facets of containment lasted throughout every administration, both Democrat and Republican, although each had a slightly different take on it.

"But wait, Professor," a student invariably interjects. "Grand strategy was possible in the Cold War because we had a clear enemy, the Soviet Union. That focuses strategy. Today we don't have an enemy, making grand strategy impossible."

"Not so," I reply. "Imperial Britain had a grand strategy that lasted four hundred and fifty years, and it did not revolve around a single enemy."

Great Britain maintained a grand strategy from the time of Queen Elizabeth I to the Suez Crisis of 1956 that endured despite the life cycles of monarchs, the various prime ministers in power, the civil and foreign wars fought by the state, the fortunes of specific political parties such as the Tories and the Whigs, and myriad other challenges. It endured because it was tethered not to a single threat, but rather to the country's geopolitical situation. Great Britain's grand strategy had five elements: transform the nation into an island fortress; gain wealth through colonization and commerce abroad; maintain naval superiority to protect sea lanes and trade; never get the army cornered on mainland Europe; and keep European rivals down by playing them off one another. This grand strategy worked into the middle of the twentieth century, when the age of colonization ended.

Grand strategy does not require an enemy to be effective. In fact, the United States arguably still has a grand strategy. It's one of hegemonic primacy, or being king of the hill in international relations. Its core elements are: remain the unilateral superpower; create a "rules-based order" in which the United States sets rules that supports its interests; shape international organizations and norms to favor US objectives; preserve military dominance and global force projection to enforce its will; maintain economic dominance (dollar as global currency); uphold cultural dominance (English as universal language, awesomeness of the entertainment industry); and promote democracy and free trade. All of these elements are undergirded by "American exceptionalism," an ideology that asserts that the United States is unique among nations with a duty to defend democracy and personal freedoms around the world.

However, American grand strategy is fading. It's being challenged by rivals like China and eroded by the inexorable forces of durable disorder. A new grand strategy is needed for such an era. What it should entail remains an open question. At a minimum, it should cleave to the New Rules of War. There are other options, too. For example, a new grand strategy could include securing pockets of order, disrupting those who hasten disorder, and maintaining freedom of action in zones of disorder. It may take the form of following multinational corporations into contested areas or zones of disorder, in a "flag follows trade" policy. The private sector is full of savvy diplomats who represent companies rather than countries. Lastly, a new grand strategy should promote the free flow of ideas and information. Truth dies in the dark.

Absent from this list is ideology. What a country stands for—democracy, human rights, values—is a vital component of its national soul, but it gets countries in trouble. The core problem is hypocrisy, and the United States is a good example. Endless National Security Strategies assert that America will defend democracy and human rights—until it doesn't. In fact, the White House routinely defends autocracies like Saudi Arabia. Such hypocrisy undermines US credibility, and credibility is power in foreign affairs. Just look at North Korea and the United Nations. The North Korean regime constantly issues threats and dictums that are met with loud yawns from the international community because the country is no longer credible. The West must avoid this path.

"Without grand strategy," I say to Dan, "we're an A-bomb of unpredictability, because even we don't know what we're doing."

Dan puts down his sweet tea, pondering this. Then he gives it one last slurp before tossing the cup into the trash.

"Predictability. Maybe that's a crock of shit, too," he says, getting up.

"More fighting words."

"You can't be too predictable, otherwise the enemy figures you out."

As we collect the rest of our trash, a squad of colonels holding sandwiches approaches us, obviously coveting our rare Pentagon park bench. The building has 23,000 people and only a dozen park benches.

"You guys leaving?" asks one of the colonels.

"Yup," Dan says.

"Great, 'cause we were going to fight you for the bench," jokes the colonel. He's a marine, so it's a credible threat.

As we leave, I turn around and shout across the courtyard.

"Hey, Colonel. Why didn't you fight us for the bench? You had the numbers!"

"Because we knew you guys were about to leave. So why fight?" he shouts back.

"See, predictability is not a crock of shit," I say to Dan. "Nor is grand strategy."

Dan grunts as we walk up three flights of stairs and turn down the Marshall Corridor. General George C. Marshall was more than a general—he was a grand strategist. Head of the army during World War II, secretary of state, secretary of defense, president of the American Red Cross, and architect of the "Marshall Plan" to reconstruct postwar Germany: the guy left no room for the rest of us.

"Yeah, maybe. But where's our Marshall today?"

"Good question," I say.

RULE 4: HEARTS AND MINDS DO NOT MATTER

First-century Judea was a pit of asps for the Romans. Jews refused to bow to foreign overlords, no matter how powerful. Rome had struggled against Jewish rebellion since Pompey conquered Judea in 63 BCE. God's law trumped Caesar's, the Jews believed, and the magnificent Temple in Jerusalem was the center of the universe, not Rome. So powerful was their faith that several proclaimed themselves the Jewish messiah, who would usher in the Kingdom of God. One of these men was Jesus of Nazareth.

Roman taxation was tantamount to slavery for the pious, and some Temple priests encouraged Jews to protest by not paying taxes. Tension reached a breaking point in 66 CE when the Roman procurator Florus ordered troops into the Temple, defiling it, to collect taxes owed from its treasury by force. Protests ignited across Jerusalem, and Jews openly mocked Florus by passing around a collection basket, as if Florus were poor. Rioters even attacked the Roman garrison, killing soldiers.

Florus reacted badly. The next day, he sent soldiers into Jerusalem to arrest city leaders, who were later whipped and crucified, even though many were Roman citizens. Outraged Jewish insurgents took up arms and overran the Roman garrison, who surrendered and were lynched. The pro-Roman king, Agrippa, fled for his life, while Judean rebels sprouted up everywhere, killing Roman sympathizers and cleansing the country of all Roman symbols.

Cestius Gallus watched the situation with alarm. He was the Roman legate in neighboring Syria and knew that if one province successfully revolted, others would follow. Soon, the entire empire could be in rebellion, and the Eternal City lacked the troops to suppress it. Better to quash dissent while it was still budding, and that's what Gallus intended to do.

Gallus assembled an army of 30,000 to 36,000 troops, with the mighty Twelfth Legion at its core. The others were "auxiliary" troops, mercenaries and foreign allied armies. Legions were the backbone of the empire, and Rome brought its known world to heel with just twenty-seven of them. Each consisted of about 5,200 elite heavy infantry, recruited exclusively from Roman citizens and drilled to perfection. Legionnaires were both feared and lionized throughout the empire.

The Roman army cut through Judea like a gladius, massacring thousands of people and rebels alike. Town after town fell, until it arrived at Jerusalem. Its walls were too thick, its defenders too ensconced, and its food stores too deep for a quick victory. Time for reinforcements, Gallus thought, and he withdrew to the coast with the Twelfth Legion and some auxiliary troops, leaving the bulk of his army to besiege the holy city.

Rebel scouts stalked the Roman forces as they snaked through the desert ravines. Gallus thought nothing of it. Small groups of rebels posed no threat, especially since he traveled with the

12th Legion. In months of campaigning, all insurgents had succumbed to this juggernaut of Roman will.

The road narrowed in a mountain pass about twenty miles northwest of Jerusalem, at a place called Beth Horon. The army slowed and became tightly packed as it squeezed through the pass. A barrage of arrows dimmed the sky and swarmed the compressed army, killing many.

"*Ad aciem! Pugna! Celeriter!*" ordered the centurions. "Form battle lines! Fight! Quickly!"

"*Rak chazak amats!*" The ancient Hebrew battle cry echoed around the ravine, drowning out the commands. Tens of thousands of heavy Judean infantry streamed down the mountainside from hidden positions, ambushing the trapped Roman army on every side.

"*Ad testudinem! Ad testudinem!*" shouted the centurions. "Form the tortoise!" Cohorts of the 12th Legion hastily interlocked shields on all sides and above their heads, creating a shell of steel.

An avalanche of Judean spears rushed down the valley walls and overran the auxiliary troops' hastily formed defensive lines. With nowhere to retreat, the survivors were pushed back and into the legion's tortoise formation, breaking it up.

"*Gladium stringe!*" "Draw your sword!" In an instant, five thousand swords were unsheathed. "*Repulsus!*" "Drive them back!"

The Judeans and legionnaires smashed into one another, heavy infantry head on head. The combat was intense. The Judeans fought as if commanded by God, while the legionnaires battled for Roman honor, their religion. Numbers won out. In the end, the 12th Legion was annihilated, and their prized eagle standard, called an *aquila*, was taken as booty. In the empire's thousand-year history, only a few *aquila* had ever been lost to enemy forces. Beth Horon was one of Rome's worst defeats.

Now it became personal for Rome. The superpower dispatched its finest general, Vespasian, with an army of 60,000 troops—including three full legions—to crush Judea. It was the world's most powerful army, bearing a grudge. The Romans quickly reconquered Galilee, collapsing the rebellion in the north. An estimated 100,000 Jews were killed or enslaved. Some were insurgents, most were not. It didn't matter in the eyes of the Romans. Abetting rebellion was a capital offense, a lesson they wished to instill across the empire.

After systematically ridding the countryside of insurgents, Vespasian turned toward the heart of the insurgency: Jerusalem and the Jewish ideology that had fueled the rebellion. Vespasian was no effete like Florus, who had achieved his position through his wife's connections. Vespasian was a self-made Roman general, the kind that built a millennial civilization. This is also why he was called back to Rome in the middle of the campaign to become emperor, ending a civil war that nearly drowned Rome in the Year of Four Emperors (69 CE).

Vespasian left his son Titus, himself a future emperor, to obliterate the Jewish insurgency. By this time, Jerusalem lay under supersiege. The Romans had built an earthen berm as high as Jerusalem's famed stone walls, surrounding the city and sealing it off from the world. Nothing got in or out. Anyone caught in the vast dry moat was crucified on the berm, facing the city for all to see. The screams of agony lasted for days, followed by weeks of stench as the bodies rotted on the cross. Up to five hundred crucifixions took place a day.

Inside the walls, life was equally bad. A civil war broke out between two Jewish factions. One side was led by Ananus ben Ananus, a former high priest. Opposing him were the Zealots and Sicarii, fanatical religious terrorists. Unwilling to consider any opinion but their own, they seized the Temple by force and commanded the Jews obey them. Ananus's militia surrounded it, creating a Jewish siege within the Roman one. Zealots used trickery to massacre Ananus, his followers, and many common people. Now in control, the extremists murdered anyone who spoke of surrender. Flavius Josephus, an eyewitness, described their rule as a reign of terror, one in which the fanatics executed dissenters using sham tribunals. The Zealots even destroyed the city's food supply so that the people would be forced to fight against the Roman siege instead of negotiating peace. All it achieved was more starvation.

The Sicarii were cloak-and-dagger assassins, predating the Islamic Hashishin ("Assassins") and the Japanese ninjas by centuries. Like the Zealots, they were terrorists. They took their name from a thin curved dagger called a "sicae." (*Sicarii* literally means "dagger men.") At public gatherings, they would furtively stab victims and then blend back into the crowd, concealing their daggers under their cloaks while bewailing the murder alongside everyone else, then slip away. Their targets were Romans, Roman sympathizers, and Jews they thought apostate. They also raided Jewish villages like Ein Gedi, where they slaughtered seven hundred women and children during Passover. In later Latin, *sicarius* became synonymous for a murderer.

When the Romans finished constructing siege towers and battering rams, they smashed through Jerusalem's city walls. The army spilled into Jerusalem, slaughtering all in their path and burning the city. Defenders made a last stand in the upper city but were overpowered. The five-hundred-year-old Second Temple—the symbol of Judaism—was desecrated, plundered, burned, and torn down stone by stone. When that was done, legionnaires turned their wrath to the Temple Mount, pushing its huge stones over the side, where they lie today at the foot of the Wailing Wall. Jews still mourn the destruction of the Temple. Once the fighting was over, the Romans killed the elderly and most military-age men, then sold the women and children into slavery. Tens of thousands were killed or enslaved that day, and Emperor Vespasian used the spoils from the Temple to pay for the Colosseum in Rome.

However, the Jewish insurgency was not dead. A thousand Sicarii terrorists escaped through hidden underground tunnels and sewers, and made their way to Masada, the ancient world's most impregnable fortress. Located on the edge of the Dead Sea, Masada was built on a 1,500-foot mesa reminiscent of an island of stone in the Grand Canyon. A few narrow paths carved out of the cliff face were the only way up the mountain, which were easy to defend with a small force. The Sicarii surprised the Roman garrison occupying Masada, killing all seven hundred of them, and took over the mountain fortress. It had enough stores to feed the Sicarii and their families for years. They would wait out the war and begin their rebellion anew once the Roman army left Judea. Or so they thought.

Who would attempt to conquer Masada but the Romans? General Silva arrived with the 10th Legion and fifteen thousand Jewish slaves from Jerusalem, and methodically set to work building seven fortified camps and a berm, surrounding Masada on all sides. Then, unimaginably, Roman engineers began to build an earthen ramp from the desert floor to the top of the mountain. It took a year to construct the giant ramp using hand tools. They also created a multi-storied siege tower with a battering ram. Somehow, in a colossal engineering feat, they hoisted it up the ramp until it scaled Masada's fortress walls. It broke through in a single day.

What happened next has become legend in Jewish culture. With nowhere to run, the Sicarii watched their impending doom in glacial slow motion, until the fateful day their fortress wall fell. The Romans returned to their camps for the night, perhaps reasoning that slaughter is easier with a good night's sleep. Inside the fortress, the Sicarii men gathered outside the synagogue to listen to their leader, Eleazar ben Yair. He motioned them to gather closer, so all could hear.

"Long ago we resolved to serve neither the Romans nor anyone else but only God," Eleazar said. "Now the time has come that bids us prove our determination by our deeds. . . . Let us die before we become slaves under our enemies, and let us go out of the world, together with our children and our wives, in a state of freedom."

These words steeled their hearts for the end. Some men sobbed, but the thought of their wives being gang-raped by soldiers while their children were taken away for a lifetime of slavery in service to pagan gods was too much. When Elazar finished speaking, all swore they would not be taken alive.

Rather than fight to the death, they opted for mass suicide. Since suicide is prohibited in Judaism, the men killed their wives and children first, then gathered in the bathhouse. Each wrote his name on a shard of pottery and threw it into a pot. They took turns drawing a name, then killing the man listed until only one remained, who either committed suicide or was killed by a woman.

When the Romans stormed up the ramp the next morning, they found 960 bodies. Only two women and five children were found alive.

After this, the Romans had no more problems in Judea for decades. The First Jewish-Roman War (66–73 CE) is an example of a successful counterinsurgency, or "COIN." Winning hearts and minds, which is how the West conceives of COIN today, is irrelevant. The Romans made a desert and called it peace, to paraphrase the ancient historian Tacitus. Bloody determination and strategic patience eradicated the roots of insurgency and won the war. It is not fair, just, or moral. But it is effective. This is what successful COIN requires, and anything less simply prolongs the fight, as demonstrated by the failure of modern COIN in Iraq and Afghanistan. Like all forms of armed conflict, COIN isn't for the kindhearted. "War is hell," explained General Sherman during the American Civil War. Sherman's March to the Sea was a "scorched earth" strategy that proved decisive for the Union's victory.

COIN OF THE REALM

In 2009, I sat in an overflow room at the Willard Hotel in Washington, DC, listening to General David Petraeus explain how the only solution for the failing war in Afghanistan was a "comprehensive counterinsurgency strategy," modeled after the one that had allegedly won Iraq.

Petraeus's speech came at the annual meeting of the Center for a New American Security, a DC-based think tank that had become a locus of COIN thinking. The four-star general was at the peak of his power; he was *Time* magazine's Person of the Year in 2009 and lauded by both Democrats and Republicans for saving Iraq. His thick PowerPoint presentation spoke of securing and serving the population, understanding local circumstances, separating irreconcilables from reconcilables, and living among the people.

This became the Petraeus Doctrine: winning hearts and minds is the only pathway to victory when fighting unconventional wars. The general and his entourage, nicknamed the COINistas, called it population-centric, or "pop-centric," COIN, and they promised it would win Afghanistan. The audience applauded and cheered. People always like good news.

Petraeus was the great savior of Iraq and Afghanistan, until they blew up. People think that saving the population wins the war, but this rarely works. War is too complex for such facile sentiments, which is why the United States lost in both Iraq and Afghanistan. Nevertheless, Petraeus and pop-centric COIN remain mainstream strategic ideas, which poses a dangerous problem. Why? Because we have created new myths about strategy that will persist for years despite their obvious failings, and we will make bad decisions about intervening in future wars based on these myths.

Years earlier, Petraeus was my brigade commander in the 82nd Airborne. Back then, in the early 1990s, he was just a colonel and I a lieutenant. I remember one night we conducted a "mass attack" in which two thousand paratroopers descended onto a mock enemy airfield in the middle of Fort Bragg at night. Our mission: capture the airfield. After we cleared the enemy, I happened to run into Colonel Petraeus on the battlefield, a gaunt figure and an exercise nut.

He asked me how my platoon had performed, and I told him.

Then we had an erudite conversation about the future of war, surrounded by smoke in the purple dawn twilight. It was surreal.

"Future wars will not be conventional," Petraeus said. "We're not going to be fighting states, or people who fight like us."

"Who, then?"

"Guerillas, drug thugs, and others who want to depose lawful governments. In a word, insurgents."

"Like Vietnam?" I said. "That didn't end well for us."

We discussed why, then he said, "You should leave the army and get a PhD."

"Why?" I asked, crestfallen. At that point, I still wanted an army career.

"Because counterinsurgency is the future, and the military isn't ready for it. But people need to be. Study it and strengthen your mind."

So I did. My battalion commander, a young Stan McChrystal, even wrote my recommendation for Harvard. Ironically, I learned that counterinsurgency rarely works. As T. E. Lawrence ("Lawrence of Arabia") explains, armies are like potted plants: strong but immobile. They are rooted to fortresses, and when they move, it's a crawl. Meanwhile, guerillas are like vapor, invisible and everywhere. They come and go as they please and are too clever to be caught in open battle with an army, which would obliterate them. Better to be vapor. You can't hit vapor, and that's their defense.

But insurgencies are more than guerillas. They are armed social movements that want to topple governments. Insurgents dream of leading their people into revolution, as Lenin and Mao did. Ideology is their weapon of choice, as they strive to win over the people's "hearts and minds," a term dating back to the American

Revolution, another insurgency. Insurgents' mission is to convince the population that the current regime is despotic and illegitimate, worthy of overthrow. The more the regime's forces crush the insurgents, the more they prove the insurgents' cause to the people. Hence, conventional war strategies guarantee defeat.

The COINistas get some things right. They espouse ditching conventional warfare when facing an insurgency. Instead, they argue, beat the insurgents at their own game, and start your own armed social movement to compete with theirs. You win by capturing more hearts and minds, not battlefield victories. "Military action is secondary to the political one," argues David Galula, a seminal COIN strategist and a French army officer. Many of the big ideas in the United States' much-hyped *Counterinsurgency Manual* in 2006 were filched from Galula's writings from the early 1960s.

In those writings, Galula lays out an eight-step strategy to win. It comes down to this: isolate the people from the insurgents, keep the area secure, earn the population's trust so they tell you where the insurgents are hiding, and then mop up the insurgents. "Mop up" is despot's jargon for "kill or imprison for life." It's a clear-hold-build campaign, which the United States adopted in Iraq and Afghanistan. However, Galula's strategy also involves rigging elections, controlling the media, neutralizing political opponents, and replacing (disappearing?) elected officials who don't agree with you. It's very Putin. Petraeus didn't do this, at least not all of it, and instead chose nation-building. That didn't end well.

But pop-centric COIN gets the big things wrong. COINistas are fond of saying that COIN is a fight for legitimacy because this wins hearts and minds. However, they are pretty dumb when it comes to what "legitimacy" comprises. When considering Iraq or Afghanistan, they mistakenly assume legitimacy is just like it is back home in the West. This is imbecilic. In a democracy, legitimacy is conferred by the people's consent to be governed—hence the importance of elections. People owe their obedience to the government in exchange for social services like security, justice, education, and health care. If the population is dissatisfied, it can fire the government and elect new leaders. Political scientists call this dynamic the "social contract" between ruler and ruled.

COINistas think you can forge a new social contract in failed states if you provide people with better social services, literally building a nation out of dust. One COINista even called it "armed social work" (which angered social workers everywhere). As a result, the United States blew billions in Iraq and Afghanistan building schools, roads, hospitals—a state. But this never succeeds because—spoiler alert—populations are not bribable. Individuals can be bribed, but not communities. It turns out that people will take your stuff but not your ideology. Would Americans become Communists if China built them new schools and hospitals? Heck no. Yet this is the logic of COINistas.

Imposing a Western concept of legitimacy in Iraq and Afghanistan was a mistake. In the world of intelligence analysis, this gaffe is called "mirroring," and it's a cardinal sin. It assumes other populations think the same way as we do, but foreign societies do not mirror our own. This would seem obvious, but not to the COIN crowd. Legitimacy in societies like Iraq and Afghanistan is not conferred by a democratic social contract but rather by political Islam. Piety to god and observance of sharia law matters most, which is what al-Qaeda, ISIS, and the Taliban are selling. Pop-centric COIN was doomed to fail because of this blunder.

COIN's imperialist origin is another reason why it flopped so spectacularly. COIN was never meant to build democracies—it was designed to enslave people. Early COIN theorists like David Galula, Roger Trinquier, C. E. Callwell, and others were European colonialists. Their goal was to impose a colonial regime, not to create independent states. Galula wasn't interested in transforming Algeria or Indochina into democracies. He wanted to reestablish the French colonial grip over locals who sought their freedom, and he devised COIN to achieve this. Rigging elections, manipulating the press, ordering extrajudicial killings, and engaging in other activities anathema to democracy were acceptable techniques because the places in question were just colonies. Trinquier advocated torture and brutality. Imperial powers establish colonies mainly to extract wealth, making COIN the wrong model for the United States in Iraq and Afghanistan.

If you begin on the wrong foot, you will continue to trip on the path ahead.

SUCCESSFUL COIN STRATEGIES

The biggest problem with COIN is that it ignores history. Insurgency scholarship generally disregards the huge number of insurgencies that fail and are consigned to the dustbin of history as flopped revolutions, rebellions, or just plain crimes. Mostly, insurgencies are flattened by government forces. It is interesting to contemplate how the weak win wars so long as one keeps in mind that most of the time they don't. Here are three COIN strategies that can succeed—if you can stomach them. None triumphed by winning hearts and minds.

The first is the "drain the swamp" strategy, and it's what the Romans did to Judea. Clausewitz liked this one, and he thought of peasants-in-arms as rabble to be put down like obstreperous curs. But Clausewitz had a point, and this is usually what happens in history. The biggest challenge is finding insurgents to kill. Before Mao was supremo of China, he was a lowly insurgent and always on the run. He used to say: "The guerrilla must move among the people as a fish swims in the sea." This means guerillas must blend in with the local population to survive, as the Sicarii did. One solution is to drain the sea or swamp, exposing the fish so you can kill them. Usually this means blasting the population until the insurgency is dead, collateral damage be damned.

Coercion has worked for many. During the Vietnam War, the US Army Green Berets had a saying: "If you grab them by the balls, their hearts and minds will follow." In 1999, Russia crushed the Chechen revolt by laying siege to Grozny, the Chechen capital, and pummeling it to dust. The United Nations declared Grozny the most destroyed city on Earth. Sri Lanka ended its twenty-six-year civil war with the Tamil Tigers after trying everything else first. In 2009, the Sri Lankan government decided to push the Tigers into the sea, and that's where they remain. Perhaps it's only genocide if you fail. Even terrorists use this strategy. Al-Qaeda, ISIS, and the Taliban built their kingdoms with coercion.

The second is the "export and relocate" strategy. During World War II, Stalin had a problem with uppity Chechens, who wanted to use the opportunity to break away from the Soviet Union. The Steel Man's answer was Operation Lentl. The Red Army forcibly spread the Chechens across the USSR's eleven time zones, so that they would be a minority in someone else's homeland, extinguishing the Chechen insurgency. Of the 496,000 people who were deported, at least a quarter perished.

The third is the "import and dilute" strategy. Shortly after Mao took power, China annexed Tibet by claiming it had once belonged to "greater" China, even though Tibet has almost no native Han people. In 1950, China's million-man army conquered the mountain kingdom in what it now calls the "peaceful liberation of Tibet." Buddhist nuns were raped, unarmed monks slaughtered, and temples looted. In the years to follow, China imported millions of Han Chinese, making Tibetans a minority in their own homeland and easier to subdue. In 2006, China opened a bullet train into Tibet, literally accelerating the process. Operating at 16,627 feet, it's the world's highest speed train, and the Chinese had to order custom locomotives to operate in the thin air. Diluting the native population with your own smothers potential insurgencies.

The best way to kill insurgencies is to use all three strategies at once. Rome ruled for a millennium this way, and it's how outnumbered European colonialists controlled rebellious territories. The United States conquered its western frontier using these techniques, to the humiliation of native American Indians. Morocco used this combination to occupy the disputed Western Sahara territory, relabeling it its "Southern Territories." Many argue that Israel does the same with Palestinians today.

In the end, effective COIN is brutal and heartless—the opposite of Petraeus's warm and fuzzy version. If the West were to undertake effective COIN, it would harken back to the nefarious days of colonialism. The alternative appears equally problematic. Terrorism and insurgencies have been on the rise since the end of the Cold War and have come to define war today. These forces of durable disorder are a genuine threat if left to fester. Western militaries and UN peacekeepers have proved inept at dealing with them.

Small changes are the enemies of big changes. The West needs a long-term presence in zones of disorder to prevent problems from becoming crises and crises from becoming conflicts. There is a solution, but it's unorthodox. Conventional war fighters will detest it, but they are the enemy of change.

MUSTER A FOREIGN LEGION

There is no substitute for boots on the ground. Troops are needed to root out the enemies in the shadows, where they breed, before they develop into full-blown insurgencies or mature into terrorists who can attack our homeland. This requires a long-term commitment to regions of disorder, but Western societies hate seeing their troops come home in body bags. "Bring the troops home!" is a common antiwar chant. Unpopular wars guarantee lost elections, so presidents and prime ministers try to keep troop numbers down. However, this is the worst strategy, because it deploys sufficient troops to get killed but not enough to win. How can the West maintain a long-term presence on the ground without risking its own troops?

Some think covert special operations forces can do the job, but they are not built or resourced for it. Their staying power is limited, their numbers too few, and the need for them too great. Even if we quadrupled their size, it wouldn't change the essential fact that they are designed for quick-strike actions and not the enduring presence needed to quell brewing insurgencies. Others think air power is the answer, as was tried in the Balkans and Libya, but you cannot hold territory from the sky.

Western countries should create foreign legions. When people think of foreign legions, they think of French mercenaries. This stereotype is incorrect. The French Foreign Legion is a part of the French military, led by French officers, and equipped by the French government. It takes its orders exclusively from Paris, and it rewards its legionnaires with French citizenship. It's a French army unit, except its enlisted ranks come from all over the world. It functions as a quick-response force, and an elite one at that, drawing largely from the veteran pool of other militaries worldwide. Even American vets have a hard time making the cut. With seven thousand legionaries, the unit can deploy deep forward in places like Africa or the Middle East to secure French national interests.

It's time for an American Foreign Legion—and a British one, an Australian one, a Danish one, and any other country that wants to overcome threats before they arrive at their borders. Like the French model, the American Foreign Legion would be a part of the Department of Defense, except its enlisted ranks could be recruited globally—a huge pool. The United States would recruit, train, sustain, and command these troops in the long term. The legion's units would be led by American officers and special forces teams, scaling their mission at a reasonable rate.

Loyalty would be ensured by welding legionnaires' long-term

interests to Washington's. Like soldiers, legionnaires would sign multiyear enlistments and could make a career in service to America. Beyond a paycheck, the legion would also offer a pathway for citizenship. This is not a radical idea. For decades, the United States has offered earned citizenship through military service. The legion would serve as a beacon for men and women who want to opt in to the American way of life and are willing to earn it.

A foreign legion could provide the United States with long-term boots on the ground in places it needs them the most, solving a perennial strategic problem. The West's aversion to troops returning in body bags would not be an issue, judging by the US public's lack of interest in dead private military contractors or proxy militia members. The transition from US casualties to non-US ones would give the legion political freedom of maneuver to bash threats where they breed, and to take some risks doing so. Even better, once the legion eradicated a threat, it would remain in the region to prevent the threat from returning. This fixed posture would solve the problem of a US playbook limited to air strikes and the involvement of special operations forces, who can linger in a threat area only for hours or days at the most. The legion could stay for years.

The foreign legion would be designed to combat the forces of disorder, and it should be deployed into zones of chaos that are critical to us. For example, it could preemptively annihilate terrorists where they nest, hunt Iranian shadow forces like the Quds Force or Hezbollah, and kill Russia's "little green men" (who supposedly do not exist, so who will miss them?). Base the legion inside threat incubators like Syria, Somalia, and Afghanistan. Asking permission from these failed governments is pointless—they are countries in name only. Anyway, which government do you call first? These countries each have more than one. Let the legion carve out a space in a disordered world.

A foreign legion would solve other problems, too, such as that of unreliable proxy militias. Currently, Washington relies on indigenous militias, mustered in short order, to fight for American interests in zones of disorder. This has been a catastrophe. In Syria, militias armed by the Pentagon fought those armed by the CIA. At other times, militias have just handed over all their US weaponry and ammunition to terrorists. Congress approved $500 million to train and equip around five thousand anti-ISIS fighters but have "only four or five" to show for its efforts, according to the general in charge. There is no accountability for proxies, other than to end the relationship.

A foreign legion would also provide an essential boost to American troop numbers. The US military simply cannot find enough Americans willing to volunteer for service when at war. During the Iraq and Afghanistan Wars, Washington had to rely on contractors to fill the ranks, and most of them were not even American. For the first time in US military history, more contractors than troops were on the ground, and this situation created heaps of problems.

Legionnaires should replace contractors and all the troubles associated with them. Training and vetting standards could be maintained in a transparent manner, unlike with barely vetted private military contractors (I know, I was in the industry). Legionnaires would be held accountable for their actions under military law, called the Uniform Code of Military Justice. Contractors face minimal accountability. If they commit a crime, like murder, they get sent home with minimal—or no—punishment. An American foreign legion would end such impunity.

Paying for the legion would be easy. It would replace private military contractors and take their budget. In 2010, during the Iraq War, the Pentagon appropriated $366 billion for contracts—that's five times the United Kingdom's entire defense budget. Also, the amount of fraud, waste, and abuse among contractors in Iraq and Afghanistan is Valhalla in scale. The legion would serve the US government first, with no shareholders to please. Additional funds could come out of the defense budget by cutting one F-35.

Unlike contemporary forces, the foreign legion would combine the punch of special operations forces with the staying power of a conventional military unit. Strategically, this would provide a lot of agility bang for the depth buck. It would give the United States a needed weapon in zones of disorder, and it would solve other problems, too, such as inept proxy militias, wily contractors, and American casualties.

A few may blanch at the idea of a foreign legion, but it is a Hobson's choice. We either keep using the same failed strategies that waste lives, trillions of dollars, and national honor. Or we make a bold change. No one wants the former, so the choice should be obvious.

RULE 5: THE BEST WEAPONS DO NOT FIRE BULLETS

In the early hours of September 4, 1981, the Soviet army secretly streamed across the Iron Curtain, seizing vast swathes of West Germany and Austria. Echelons of troops, tanks, and aircraft rolled through the Fulda Gap on the East German border, catching NATO by total surprise. The Soviet attack was helped by a preemptive tactical nuclear strike that wiped out a quarter-million NATO personnel.

NATO defenses crumbled. Soviet armor punched deep behind NATO's front line, severing supply lines and destroying its tactical nuclear weapons. NATO's new Abrams and Leopard II tanks were outflanked by the Soviets' Operational Maneuver Group, a large task force of heavy T-72 tanks that practiced precision blitzkrieg tactics.

The Soviets used new weapons, too, including the RSD-10 nuclear missile. This fifty-four-foot weapon sat atop a massive six-axle carrier that could traverse almost any terrain, making it difficult for NATO to track all of them. The RSD-10 could deliver a one-megaton nuclear warhead 3,400 miles and had one purpose: destroy every major airfield in western Europe of use to NATO air forces. No airfield was safe, from Heathrow to Frankfurt. The USSR had 654 such missiles.

It was over within eight days. The Red Army's sheer numbers and speed collapsed NATO's forces before they could mount a comprehensive defense. The initial assault force of 150,000 Red Army soldiers was followed by millions more. Communist flags flew over European capitals, while NATO allies bickered among themselves in a secret bunker outside Brussels. The Soviet Union and Warsaw Pact countries had conquered the West.

Or so the USSR pretended in a massive military exercise designed to strike mortal dread into NATO, which it did. Operation Zapad-81 ("West-81") was the largest military exercise ever conducted by the USSR. This eight-day show of force involved between 100,000 and 150,000 troops massed along the NATO border, and it coordinated thousands of tanks, airplanes, missile launchers, amphibious landing ships, and anything else needed for a megawar. This mock full-scale invasion was meant to destabilize Western Europe and keep the Soviet republics in line. Nothing gets people's attention like the threat of force.

THE DECLINING UTILITY OF FORCE

Zapad-81 is so yesteryear. Political power used to come out of the barrel of a gun, and despots from Mao to Stalin relied on the assurance of steel to persuade the opposition of their righteousness. If the Soviets wanted to send a message to NATO, it usually involved munitions.

No longer. Today, when Russia wants to destabilize Europe, it does not threaten military action, as the USSR did. Instead, it bombs Syria. This tactic drove tens of thousands of refugees into Europe and exacerbated the migrant crisis, instigating Brexit and stoking antiestablishment politics across the continent. European law stipulates that all countries must absorb some refugees, whether they want to or not, but the law never imagined millions flooding the border. Germany alone took in nearly a million asylum seekers, requiring $6.7 billion in resettling costs. Other countries took in fewer, but the numbers still exceeded what local populations would tolerate.

Political backlash was swift, pushing the continent into its deepest crisis since World War II. Right-wing nationalist parties, once shunned as neo-Nazis, became popular in Austria, Germany, and Italy for the first time since the 1930s. Their anti-immigration stance and Euroskeptic views gained them voters across the continent. In France, Marine Le Pen almost won the presidency for the Front National, a far-right party. In Hungary, the ultranationalist "Movement for a Better Hungary" became the country's third-largest party. The United Kingdom even voted to leave the European Union, in a move known as Brexit, and others may follow its lead. The deluge of refugees empowered Moscow-friendly Euroskeptics across the continent, weakening NATO and the pan-European dream.

Putin achieved what the Soviets could not by weaponizing refugees rather than threatening firepower. This became a crisis for NATO. General Philip Breedlove, the supreme allied commander of NATO and the head of the US European Command, said Russia and Syria had turned migration into a weapon by systematically bombarding civilian centers. "Together, Russia and the Assad regime are deliberately weaponizing migration in an attempt to overwhelm European structures and break European resolve," Breedlove told the Senate Armed Services Committee. Worse, he continued, terrorists like ISIS have exploited the refugee crisis to infiltrate Europe, using fake passports that are "virtually impossible" to detect. Old Soviet wargames never achieved so much.

War has moved beyond lethality. Today, *all* instruments of national power must be used, not just the ones that shoot. Nonkinetic weapons can be very effective in war, and cunning strategists can weaponize almost anything, including refugee waves. This causes cognitive dissonance for conventional warriors, who place their faith in firepower, a concept they call the "utility of force." In doing so, they're talking about the effectiveness of violence in conflict, and they rate it supreme. For them, the application of enough force can solve any problem. Such thinking led to the meat grinder of World War I and carpet-bombing during World War II. Today, conventional militaries measure themselves by firepower. They obsess over "force projection" and "combat power overmatch," investing in things like the F-35.

Force is America's favorite tool of foreign policy, judging by where it puts its money. Budgets are moral documents because they do not lie. The United States spends twelve times more on its military than it does on diplomacy and foreign assistance. The Department of Defense's annual eleven-digit budget dwarfs that of all other governmental departments. Defenders of this staggering budget note that it is a smaller percentage of the United States' GDP now than in the past, as if this somehow represented value. This is a silly justification. America expends more on national defense than China, Russia, Saudi Arabia, the United Kingdom,

India, France, and Japan *combined*. It's overkill, but force is the oldest anthem of war.

The utility of force is declining today. Soon it will be nearly gone. Warfare has changed radically since World War II, rendering force ever-more obsolete—a curious concept that conventional warriors have yet to comprehend. The evidence is stark, though, as weapons designed with a high utility of force—fighter jets, tanks, and submarines—sit in hangars, wait in motor pools, and cruise the ocean, not fighting. Yet war has not decreased. What has changed is the role of force in modern warfare. Evidence for this is abundant, as Davids overcome Goliaths in the Middle East, Asia, and Africa.

The declining importance of force is a trend that will continue, one that will render big militaries unnecessary in future wars. This bewilders the rank and file. They will admit that weapons like stealth bombers, frigates, and nuclear missiles have been peripheral in wars since 1945, but they insist they are still needed for deterrence. This is a specious argument. Contemporary and future threats are not conquering states but failing ones, and what emanates from them are terrorists, rogue regimes, criminal empires, or just plain anarchy. None of these things are "deterrable," a fact repeatedly proved since the end of the Cold War, as terrorism and failed states have reached epic proportions despite superweapon deterrents. Traditional deterrence is obsolete.

Others argue that weapons with a high utility of force are needed for future conventional wars. This is another spurious argument. Even if there is a big interstate war, such as one between the United States and China, why do people assume it will be fought like World War II? Conventional war is dead. There are new rules for war, and those who adapt last will die. There are other ways to win, and they do not involve bullets.

WARRIORS OF THE MIND

"Hello, I'm a Mac."

"And I'm a PC."

In this commercial, two actors appear onscreen against a white background. One is pudgy and wears a tie. He represents the PC computer and is arrogant, slow, boring, passive-aggressive, unreasonable, and, most important, a nerd. On the other hand, the Mac emanates cool. He's relaxed, wearing only a T-shirt and jeans, and displays caring, humility, and an upbeat vibe. Everyone likes the Mac.

In "Restarting," Mac and PC introduce themselves, then PC freezes. When he restarts, he goes through his "startup" introduction again, word for word. Mac explains he doesn't need to reintroduce himself, they've moved on—then PC freezes again. "We had him, then we lost him," says Mac, who asks the audience to keep an eye on PC while he calls IT for help.

Love or hate 'em, Apple's "Get a Mac" ad campaign made a huge impact. In 2006, faced with slumping sales, Apple rolled out this advertisement campaign. Three years and sixty-six commercials later, the company had tripled its computer sales and won the grand Effie Award, the Oscar of advertising. *Adweek* called it the best ad campaign of the decade, and "Get a Mac" remains iconic to this day.

How did Apple do it? The secret is simple: denigration. Going negative is powerful, but the trick is to make the target look like the wrongdoer. Viewers identify with the cool Mac, as opposed to the stuffy PC. However, the Mac is not a nice guy. He eviscerates the PC, getting the audience to laugh as he does so. The Mac is the true villain, yet somehow he frames the PC as the bad guy. It's beautiful ridicule, highly manipulative, and it works.

In the future, we will need "Get a Mac" for war. Take terrorism. You can kill as many terrorists as you want, but it will not destroy the virulent ideology that spawns it. To eliminate jihadism, for example, you need to delegitimize the ideology, and ridicule does this. ISIS and its successors would shrivel like the Wizard of Oz if the Muslim world could belly laugh over them. North Korea has already been stung by ridicule's quill in movies like *Team America*, and Putin's cult of personality would whither under the power of denigration. In fact, he's easy pickings, given his naked bear-riding habit.

In modern warfare, influence is more potent than bullets. The West is terrible at it. Like the old cartoon character Mr. Magoo, it bumbles along with its ultrastealth bombers and its *Leave It to Beaver* strategic communications skills. It still tries to change minds using Cold War techniques like Radio Free Europe and dropping pamphlets out of the sky (something the air force calls "bullshit bombs"). We just look like the uncool PC.

Weaponizing influence and controlling the narrative of the conflict will help us win future wars. The United States should be winning the battle of narrative—it's the home of Hollywood and Madison Avenue. Yet it's been lapped by Russia, Iran, and China. Terrorists are exceptionally good at weaponizing information. "We're being out-communicated by a guy in a cave," Robert Gates, the secretary of defense, used to say about Osama bin Laden, and nothing has changed in the years since his death. This is Washington's conclusion, too, according to internal government reports.

The West needs to update its information-warfare game. Until it does, it will continue to get outplayed by its enemies that wage war in the information space, and that's most everyone. In America's case, this will require structural change. Currently, no one in Washington really knows who's in charge of strategic influence. Is it the State Department, the military, the CIA, the National Security Council, or something else? Yes, they say. No wonder the superpower is losing. The correct answer is the CIA, because only it is authorized to conduct covert, or "Title 50," programs, which are essential for this kind of warfare. But the CIA should just manage it, because bureaucrats are not artists. Instead, it should outsource the heavy lifting to Hollywood and invest real money. The Pentagon spends $120 million on a single F-35 that never flies in combat—surely some money can be spent on something that might be useful in war.

The United States is also hampered by Cold War relics like the Smith-Mundt Act. This law prevents the country from directing propaganda to its own citizens. When it was written—in 1948—it was possible to isolate Americans from propaganda targeting foreigners. Now it's not. Today we have the internet, satellite TV, smart phones, email, and other information channels, making it impossible to tune the world out. As with conventional war, Smith-Mundt and laws like it have exceeded their expiration date. Those who want to win future wars yet insist such changes are too difficult are just throwing young soldiers' lives at the problem.

Weaponizing influence and gaining information superiority has three components. First is monitoring: intelligence agencies identify who is messaging what to whom, along with how and why. As Sun Tzu counsels, know your enemy. Second is discrediting: pinpointing fake news, alternative facts, bots, trolls, false narratives, viral memes, and negative frames, and then exposing them. Mythbusting must happen, otherwise people may start to believe the spin. This task is especially critical for democracies, since enemies target Americans and others who get to choose their leaders. Third is counterattacking, and this is where Western countries grow weak in the legs. The United States and others already find ways to secretly support friendly voices in foreign lands whose messages are music to our ears; sometimes we even do this without them knowing we're helping them. This "big mouth" stratagem is good for countermessaging, but it is not enough. We need more weapons in our influence arsenal, and here are a few:

- *Denigration.* Going negative is powerful, but it must be done artfully. You need to look like the good guy while shredding the target. The "Get a Mac" campaign does this and has much to teach us: tone may be more important than information; frame the target as the aggressor using cultural "irritators" known to the audience; convey empathy and goodwill for credibility; show humility; leverage the audience's wishes to identify with a protagonist of your own choosing; align with preconceived knowledge; be funny but not stupid. You don't need to win over the entire audience, just an active minority and enough opinion leaders to seed doubt. This will demoralize, embarrass, and create distrust of targeted individuals and institutions. This works especially well against autocracies because they are often built on a cult of personality and the infallibility of leadership. Make such leaders fallible.

- *Involuntary Internalization.* Who can forget shows like *American Idol*? This singing competition has lasted more than fifteen seasons, is broadcast to over one hundred nations, and has given rise to imitators across the world. The appeal of *American Idol* is universal because it promotes a merit-based Cinderella story: anyone with enough talent can become a superstar. There are judges, but it's the audience who decides the ultimate winners by voting with their phone or online—sort of like democracy. The US should covertly sponsor or import "Idol" shows in countries with repressive regimes like Iran, Turkmenistan, and Saudi Arabia, seeding the tenets of democracy in peoples unaccustomed to it. This is possible in closed societies because it is increasingly difficult to shut the internet out of national borders. Through analogous action and associative reasoning, people may begin yearning for "voter" participation in their political life and start questioning authority. Entertainment shows are more effective than dreary exported news channels like Radio Free Europe that sound like infomercials or, worse, propaganda. People like to have fun.

- *Velvet Regime Change.* "Together, we are many! We cannot be defeated!" chanted tens of thousands packed into Kiev's Independence Square. Neither the biting winter nor the secret police deterred them, as they marched waving orange flags and demanding democracy. The "Orange Revolution" peacefully toppled the dictatorial government and inspired other "color revolutions" around the world. Authoritarian regimes including Russia and China fear them, and they blame the West for covertly orchestrating such events. Were that true. Most color revolutions failed, but they suggest a blueprint for velvet regime change—with some covert help.

- *Moral Corruption.* Socrates was killed for corrupting the youth, and so was Homer Simpson. At least that's what Iranian officials tried to do, citing the Springfield family's notoriously self-centered and irreligious attitudes as offensive in their fight against "Western intoxication" (d'oh!). But Homer is not alone. Barbie dolls were also banned, but no one knows why. Perhaps because she's busty and blond. If Homer and Barbie are threats to the Iranian regime, then why doesn't the West find creative ways to get them into the hands, hearts, and minds of young Iranians? A colleague in the intelligence community once told me we could probably shorten the Taliban's fighting season if we broadcast *Baywatch* in Afghanistan. Also, apparently most jihadis are porn dogs, including Osama bin Laden. Surely a clever strategist can do something with this?

Shaping people's perception of reality is more powerful than mobilizing a carrier strike group. It can topple governments, undermine national unity, and weaken resolve in wars. Who cares about the sword when you can influence the hand that wields it?

RULE 6: MERCENARIES WILL RETURN

"You were a mercenary? Did you kill anyone?" I get this question a lot.

"I really can't say," I answer, and it's the truth. A lot of what I did was secret, or, as we say in the industry, a "zero-footprint operation."

When you need something absolutely, positively done in war, you call the private sector. Missions once conducted by special operations forces or the CIA are now outsourced; I know because I did them. I dealt with warlords, built armies, rode with armed groups in the Sahara, conducted strategic reconnaissance in hostile territory, transacted arms deals in Eastern Europe, and helped prevent a genocide in Africa.

In 2004, I received a call from a frantic client. At the time, I was in Liberia, raising its army from scratch.

"McFate?" said the voice on the other end of my satellite phone.

"This is an unsecure line," I said.

"Roger. We don't have time for that now. I need you on the first flight to—" Gunfire erupted behind me, drowning out his last words. Troops scattered everywhere. We were on the firing range zeroing our AK-47s, and someone had just blown a full magazine in automatic mode.

"Hey, knock that shit out!" I yelled to the sergeant major, gesturing at the kid with the trigger finger and the big grin. He was probably eighteen, and a few teeth short of a smile. The sergeant major, an Aussie, stomped over and unloaded a full verbal barrage on the kid, who withered with each invective.

"Say again," I said, turning back to the caller.

"Burundi."

"Where?"

"Brava, Uniform, Romeo," he said, spelling out "Burundi" using the military phonetic alphabet. "Call me when you get there. All will be explained."

Burundi? I thought. *Where the hell is that?* "Do I need to pack anything special?"

"Negative. No weapons. We're loading your card with thirty grand." Earlier, they'd issued me a credit card for expenses. "Just don't tell anyone where you're going or why you're there."

"Roger." Hell, I didn't even know that.

"And leave ASAP." He hung up.

Days later I was in Burundi, a small speck in the middle of Africa and one of the most dangerous places on earth. That evening I was sipping Coke with the president inside his armed palace. The room was pure African kitsch, filled with grotesque tribal masks, overstuffed leather couches with fake ebony trim, and an imitation zebra rug.

Joining us was the American ambassador, the CIA chief of station, and the president's eight-year-old daughter. They had just given me my mission brief, and now we were sipping Cokes. That's how the worst briefs end: stunned silence with a beverage.

It turns out I was not the first person in this situation. Initially, the ambassador had sent a CIA team from the ground division, but they had left after declaring it mission impossible. Then a US special forces unit had shown up, who'd reached a similar conclusion. On the way out, they'd joked that the Department of Defense doesn't do Africa. It wasn't a good joke.

That's when they turned to my employer, a large private military company who never said no to a job. I was their man in Africa, so the job fell to me. As a young paratrooper, I'd learned that nothing is impossible for the man who doesn't have to do it.

The mission was pretty impossible: I had to stop a genocide before it started. Worse, I had to do it without anyone—not even the president's men or the US embassy's staff—knowing what was happening. Plausible deniability is one of the main appeals of contractors. If something bad happened to me, I could be disavowed. If a CIA or special operations team got in trouble, the US government would have to do something: stage a rescue, pay a big ransom, or—worse—go public. But not so with contractors, who are disposable humans.

"Keep the president alive," the CIA officer told me during the brief. "He's the key."

"How so?" I asked. Unlike in the movies, no one hands you a classified dossier with all the background in real life, at least not in the private sector. There, the unofficial motto is "Figure it out."

"Remember the Rwandan genocide of 1994?" Rwanda and Burundi share a border, and the genocide scorched both countries. "Eight hundred thousand killed in three months."

"By machete, mostly," added the ambassador.

"It started when the presidents of Rwanda and Burundi were assassinated."

"Wait," I said. "How does killing two presidents lead to a genocide?"

"Because we live in a dangerous world," said the president impatiently. "Hutus and Tutsis have been at each other's throats for decades. When a Tutsi is murdered by Hutu, they kill three Hutus in revenge. Then the Hutus kill six Tutsis, and so it goes." His hands were quivering in anger, sloshing Coke on the fake zebra rug.

"And that's how a genocide begins," said the ambassador.

I looked to the eight-year-old girl, but she was watching television.

"So, you can imagine what will happen to the Tutsis if a Hutu president is assassinated," said the ambassador, eyeing the president.

Genocide, I thought.

"And that's why you're here," said the CIA officer. "We have all-source intelligence that the FNL . . ."

"Who's the FNL?" I asked.

The president looked displeased.

"The Forces Nationales de Libération, or FNL," answered the ambassador. "A nasty lot. Hutu extremists currently hiding in the jungles of Eastern Congo."

"Only twenty kilometers from here," said the CIA officer. "They want to ignite a genocide by assassinating the president."

"Why? Also, they're Hutu. Isn't he . . . aren't you," I looked at the president, "Hutu too?"

"Murders, fanatics, extremists," said the president. "They do not care. They only want to see all Tutsi killed, forever. What do they care about collateral damage?" He eyed his daughter, still watching TV.

"Which is why," the CIA officer cut in, "you need to thwart the assassination attempt on the president's life."

"Do we know when they might launch an attack?" I asked.

"Weeks."

"And do we know where they are now?"

"No. Not really."

The vagary bothered me. I would need a lot more information to make this mission work. "Do we know anything about them, other than the planned genocide?"

Stunned silence followed. I sipped my Coke.

The next day, I met with Burundi's top general, who gave me a tour of the country's special forces barracks so I could inspect the soldiers at my disposal. The situation was FUBAR. They were not soldiers but hoodlums. To my eyes, they had more attitude than combat cool, and their equipment was either broken or belonged in a museum. Worse, the "soldiers" were divided between Hutu and Tutsi factions, and I could not tell who was who. I thanked the general politely and returned to my hotel to pack.

This truly is mission impossible, I thought. *I need to leave this continent and get back to my life.* At the time, I was a graduate student at Harvard taking some time off from my studies to fight in Africa—something I did not mention when talking with the academic dean. Additionally, the lack of old people in my profession bothered me, and the pay was crap for what I was doing. It was time to go.

Around midnight I awoke to a sharp knock on the door. It was the general's men, armed. They beckoned me to follow them, and I did, wondering if I was destined to become another dead statistic in the African countryside. They escorted me to the bar, cleared out by the general's presence. He was nursing a scotch. I ordered a triple.

"Good evening, General," I said, slugging down half my whisky.

Silence. Not good.

"I hear you are leaving," he said.

How did he hear that? I wondered. What could I say? *You're fucked? There's going to be another genocide and you will all die?* Better to leave now, when I still could. I just nodded, staring straight ahead.

"You know, I have nothing," he said. "My entire family was murdered in the genocide ten years ago. I could live well in Paris, with the rest of my friends, but instead I live here, in poverty. I don't own a TV, car, or house. I have only one possession: an idea called Burundi. But look at you, with your fancy watch and satellite phone. I live in poverty but you are the poor one because you don't believe in anything." He wished me luck on my exams at Harvard and left in his convoy.

That was the turning point. I returned to my room and unpacked.

The next day I started planning the mission and assembled my team. A few weeks later, the FNL attacked, and there was a night battle in the streets of Bujumbura, Burundi's capital. The president survived, and the FNL retreated back into the Congo. We averted the genocide. But that's another story.

WAR DOGS

Mercenaries are back. I know, because I was one. Today we call them different things—private military companies, private security contractors, or just contractors—but it's all euphemism. If you are an armed civilian paid to do military things in a foreign conflict zone, you're a mercenary. Few will admit it, and some companies that provide mercenary services will threaten to sue if you mention the *M* word, but it's true. Mercenaries differ from soldiers in that they fight primarily for profit rather than politics or patriotism.

People do not like mercenaries, something I know from personal experience. While at Harvard, I recall being lambasted as "morally promiscuous" in a graduate seminar, in a tone implying I sucked marrow from baby bones for fun. My old paratrooper buddies scowled that I had "sold out," "gone mercenary," and was lost to "the dark side." Yet I am proud of the work I did in the field; most of it, at least. Can soldiers really say any different?

We are taught to despise mercenaries as villains and praise soldiers as heroes, but this is bigotry. Such stereotypes are ahistorical, and there is plenty of evidence showing that both have committed noble and abhorrent acts in the past. For example, take the Iraq War. In 2007, a squad of Blackwater mercenaries killed seventeen civilians at Nisour Square, a traffic intersection in Baghdad. It sparked an international uproar, multiple high-level investigations, and constituted a strategic setback for US efforts in that country. For Americans, Nisour Square was a stain on their country's moral character, and a low point of the war. For Iraqis, military contractors look like US soldiers, and Blackwater's callous disregard for human life seemed emblematic of America's mishandling of the war as a whole.

Nisour Square is often cited as one of the conflict's worst war crimes, but who remembers Haditha? There, in 2005, in an episode similar to the My Lai massacre of the Vietnam War, a squad of marines murdered twenty-six civilians in a revenge-killing spree after two of their comrades were killed. The victims ranged from age three to seventy-six. Many were shot multiple times at close range while unarmed, some still in their pajamas. One was in a wheelchair, and four were children.

For some reason, Haditha was overlooked as an acceptable war tragedy. The only investigation—conducted by the military on the military—found nothing wrong. Incredibly, it blamed the whole thing on "an unscrupulous enemy," and dismissed it as a "case study" that illustrates "how simple failures can lead to disastrous results." After massacring twenty-six civilians, more than who died in Nisour Square, the military quietly dropped all charges against the marines except for the squad leader, who was acquitted in a court-martial. The world yawned.

Both Nisour and Haditha were comparable crimes, but people's reaction could not have been more different: mercenaries are butchers, while soldiers make innocent mistakes. This is an irrational prejudice. Murder is murder, no matter what kind of warfighter pulls the trigger. When I point this out, some people grow hostile with the burn of cognitive dissonance. Even enlightened minds balk, so strong is the bias against private warriors.

Now that I'm out of the industry, I'm often asked to talk about it in front of large audiences. My best questions come from general audiences, perhaps because they have little received wisdom on the topic and therefore a more open mind. When I speak to expert audiences—those in think tanks, universities, the Pentagon, the British House of Commons—I run into strong prejudices against using private force.

Invariably, someone throws Machiavelli at me to substantiate the mercenary stereotype. Scholars have little understanding of mercenary warfare, and what little they do stems from Niccolò Machiavelli (1469–1527), a senior official in the city-state of Florence and the author of *The Prince*, a treacherous little handbook about power. In it, Machiavelli curses mercenaries as "disunited, ambitious, without discipline, unfaithful; gallant among friends, vile among enemies; no fear of God, no faith with men." This judgment has ossified into orthodoxy.

Experts think Machiavelli's assessment of mercenaries is definitive, but it's bunk. He hated mercenaries not because they were faithless but because they cheated him, owing partly to his own ineptitudes. From 1498 to 1506, he helped organize Florence's defenses and suffered serial humiliations at the hands of his own mercenaries. During Florence's war with Pisa, a weaker foe, ten mercenary captains defected to the enemy side, a major embarrassment and a strategic blow. Amateurs should not be in charge of mercenaries any more than penny-stock traders should dabble in hedge funds.

No longer trusting mercenaries, Machiavelli convinced the Florentine authorities to raise a militia instead, one composed of citizen-soldiers whose loyalty to the state would remain unflappable. But loyalty is a poor substitute for skill. These farmers turned soldiers were no match for professional troops, and the Florentines were crushed in 1512. This fiasco dealt a death blow to the Florentine Republic, which was placed under papal control, and undermines Machiavelli's claims about the superiority of militias over mercenaries. The French regarded the flush Florentines as the epitome of military incompetence. Cheekily, Machiavelli wrote *The Prince* to impress the conquerors of Florence and win back his old job—they must have laughed.

Machiavelli's ideas on militias and mercenaries were spurned for centuries because they were terrible. Mercenaries remained the main instrument of war for the next two hundred years, and no one dared rely on feeble militias. Despite Machiavelli's protestations, the mercenary profession was considered a legitimate trade, and often the lesser sons of nobility sought careers as mercenary captains. There was no taboo against hiring private armies. It was considered no different from employing a mason to repair one's castle walls or commissioning an artist to paint one's banquet hall. Mercenaries were how wars were fought.

The disloyalty of mercenaries is also a Machiavellian trope. He ignores inconvenient facts, like Sir John Hawkwood, who was one of the greatest mercenary captains of the age and monogamous to Florence for two decades, until his death. The city honored his faithfulness with a funerary monument at the famed Basilica di Santa Maria del Fiore, which you can still see. Meanwhile, Machiavelli faded into obscurity. He is lionized today, thanks to twentieth-century scholarship, but his views on mercenaries are spurious.

People view soldiers like wives and mercenaries like prostitutes, as people who turn love into a transaction. But in my experience, every soldier has a little mercenary in him, and vice versa. When

I was in the army, I saw lots of troops reenlist for big bonuses, a transactional practice common in most militaries. For example, the US Army pays up to $90,000 for soldiers to reenlist, enough money to make mercenaries salivate. I've also seen mercenaries refuse jobs on political grounds. Some American hired guns will never take money from Russia, China, Iran, or a terrorist group; America's enemies are their enemies. The line between soldier and mercenary is hazier than most think.

The taboo against mercenaries is an invention of the Westphalian Order. Before 1648, mercenaries were considered an honorable albeit bloody trade and were a feature—often the main feature—of war. The word "mercenary" comes from the Latin *merces* ("wages" or "pay") and is no different than the *solde*, or "pay due to fighters," from which the English word "soldier" is derived. For most of history, mercenaries and soldiers were synonymous.

THE SECOND-OLDEST PROFESSION

Most of military history is privatized, and mercenaries are as old as war itself. The reason is simple: renting force is cheaper than owning it. Maintaining a permanent military seems normal today, but it's not. Paying for one's own armed forces is ruinously expensive, similar to owning a private jet versus buying a plane ticket when you need it. Mercenaries are more economical, and that's why they have existed throughout history, with today's national armies as the exception. Put another way, if you were fighting for your life and could go to war with five thousand rented mercenaries or one thousand owned soldiers, which would you choose? Especially if your enemy had five thousand mercenaries? Some, like Machiavelli, chose their own soldiers, and they were duly crushed. Most went with mercenaries.

Mercenaries are everywhere in military history, starting with the Bible. The Old Testament mentions hired warriors several times, and never with reproach. Everyone used them. King Shulgi of Ur had a mercenary army (2094–2047 BCE); Xenophon had a huge army of Greek mercenaries, known as the Ten Thousand (401–399 BCE); and Carthage relied on mercenary armies in the Punic Wars against Rome (264–146 BCE), including Hannibal's sixty-thousand-strong army, which marched elephants over the Alps to attack Rome from the north. When Alexander the Great invaded Asia in 334 BCE, his army included five thousand foreign mercenaries, and the Persian army he faced contained ten thousand Greeks. Rome used mercenaries throughout its thousand-year reign, and Julius Caesar was saved at Alesia by mounted German mercenaries in his war against Vercingetorix in Gaul. Employing mercenaries was common in antiquity.

The Middle Ages were a mercenary's heyday. Nearly half of William the Conqueror's army in the eleventh century was made up of hired swords, as he could not afford a large standing army, and there were not enough nobles and knights to accomplish the Norman conquest of England. King Henry II of England engaged mercenaries to suppress the great rebellion of 1171–1174, because their loyalty lay with their paymaster rather than with the ideals of the revolt. In Egypt and Syria, the Mamluk Sultanate (1250–1517) was a regime of mercenary slaves who had been converted to Islam. From the late tenth to the early fifteenth centuries, Byzantine emperors surrounded themselves with Norse mercenaries, the Varangian Guard, who were known for their fierce loyalty, prowess with the battle-ax, and ability to swill vast tankards of brew.

Medieval Europe was a hot conflict market, and mercenaries were how wars were fought. Kings, city-states, wealthy families, the church—anyone rich enough—could hire an army to wage war for whatever reason he wanted, and people did. Wars were fought for honor, survival, god, vendetta, theft, or amusement. Even Sir Thomas More, the great humanist and the author of *Utopia*, in which he coined the word, advocated using mercenaries to protect his ideal utopian republic.

Popes hired mercenaries, using them to obliterate enemies and purge infidels. In 1209, Pope Innocent III launched a crusade against the Cathars, a heretical sect in southern France, that would look like a war of terror today. When his mostly mercenary army stormed the city of Béziers, both orthodox and heretical Christians fled into the local church for sanctuary. The papal legate in charge ordered the army to seal and burn the building, allegedly saying: "Kill them all, God will know his own." The Holy See still uses a Swiss guard, once a fearsome mercenary unit but now part of the Swiss army, complete with halberds and tights.

Mercenaries began to fade four hundred years ago as states and their national armies gradually monopolized the market for force, one of the hallmarks of the Westphalian Order. Soldiers for hire were finally driven underground by the 1850s. Occasionally they would appear as shadowy figures in twentieth-century bush wars, fighting in the Congo and other conflict backwaters. Everyone thought them an anachronism. Until recently.

CONTRACTING: THE NEW AMERICAN WAY OF WAR

Mercenaries are back, thanks to the chum-slick of American war contracts in Iraq and Afghanistan. The large number of armed contractors in those wars has defibrillated the profession, which is surprising if you think about it. One would have thought that the world's lone superpower didn't need guns for hire, but, as with everything else in Iraq and Afghanistan, the use of mercenaries wasn't planned. It just happened.

Over half of all military personnel in the recent Iraq and Afghanistan Wars were contractors. America now fights by contract, but it wasn't always so. During World War II, only 10 percent of the armed forces were contracted. This proportion leapt to 50 percent in Iraq and 70 percent in Afghanistan. For every American soldier in these wars, there was at least one contractor—a 1:1 ratio. Often it was closer to 3:1.

Aside from unleashing a tsunami of contractors, the eruption of these wars also washed in a new breed of contractor: private military companies. They perform tasks once thought to be inherently governmental, such as raising foreign armies and engaging in combat. In other words, they are mercenaries. However, they are not like the lone mercenaries of the past, wandering the Congo during the 1960s in search of work. This new breed are multimillion dollar corporations with operations spanning the globe. They are even traded on Wall Street and listed on the New York Stock exchange, bringing war profiteering to new levels.

Corporate combatants made up about 15 percent of all contractors in Iraq and Afghanistan, but don't let the small numbers fool you. Their failures have an outsize impact on US strategy. Blackwater's actions at Nisour Square provoked a firestorm in Iraq and at home, marking one of the nadirs of that war.

Foreign audiences often ask me if the United States will outsource 80 to 90 percent of its future wars, and the rising trend lines indicate that it will. Certainly Erik Prince, the founder of Blackwater, thinks it should. He pushed a plan to replace all American troops in Afghanistan with contractors. In other words, he advocated for fully privatizing the war in Afghanistan, using 100 percent guns for hire. Invoking neocolonialism, he called for an American "viceroy" backed by a mercenary army to pacify the country. The plan would pay for itself, he promised, by exploiting mineral deposits worth "trillions of dollars" in Afghanistan. If this sounds familiar, it should. Dozens of "experts" assured American taxpayers that Iraq's oil reserves would guarantee a cost-free war. Instead, the war cost trillions. Despite this, many in Washington have found Prince's proposal appealing.

Why did the United States outsource at all? Surely the world's sole superpower could handle two rogue countries; after all, it's not like Iraq and Afghanistan were superpowers. Or at least, that's what conventional warriors thought. In the rush to war, the White House assumed the fighting in these backward countries would prove short conflicts. "Five days or five weeks or five months, but it certainly isn't going to last any longer than that," said Donald Rumsfeld, the secretary of defense in 2002. The military can "do the job and finish it fast." Nearly twenty years later, America is still entangled in both places, unwilling to admit defeat but unable to declare victory.

When these wars did not end in mere months, the White

House faced a crisis. The United States' all-volunteer military found it could not recruit enough Americans to sustain two long wars. Policy makers faced three terrible options: First, withdraw and concede the fight to the terrorists (unthinkable). Second, institute a Vietnam-like draft to fill the ranks (political suicide). Third, hire contractors to fill the ranks. Not surprisingly, the Bush, Obama, and Trump administrations opted for contractors. They now surpass the number of US uniformed troops in war zones.

These wars also marked the first time in history that corporate casualties outweighed military losses on America's battlefields. Researchers found that contractors working for the Department of Defense were between 1.8 and 4.5 times more likely to be killed than their military counterparts. By 2010, more contractors were being killed in combat than troops. And blood is not the only way contractors pay for America's wars. A RAND study revealed that 25 percent of contractors met criteria for PTSD, exceeding rates of US veterans. Additionally, 47 percent of contractors met criteria for alcohol misuse and 18 percent met criteria for depression. However, the true number of wounded and killed contractors remains unknown. The government does not track this data, and companies underreport it because it's bad for business. Perhaps this is by design. Injured contractors save the government money because it doesn't have to pay their expensive hospital bills or provide veteran's benefits, as it does for injured soldiers. From the client's perspective, contractors are disposable people who get what they deserve.

Contractors are also cheaper than the military. The Congressional Budget Office, a watchdog agency, found that an infantry battalion at war costs $110 million a year, while a comparable private military unit totals $99 million. In peacetime, the costs savings are even greater; the infantry unit costs $60 million, and the contractors cost nothing, since their contract would be terminated. From 1995 to 1997, the mercenary firm Executive Outcomes was paid $1.2 million a month to put down a rebellion in Sierra Leone—which it did—whereas UN forces swallowed up $47 million a month doing nothing. Business excels at efficiency compared with the public sector.

This disparity in cost has led to major investments in contract warfare, making war even bigger business. The market for force's value remains unknown, since there is no Bureau of Labor and Statistics for mercenaries. The Department of Defense spent about $160 billion on private security contractors from 2007 to 2012, almost four times the United Kingdom's entire defense budget. But this statistic entails only military contracts and does not include those made by other government agencies. For example, the State Department also hires private military contractors, including Blackwater, Triple Canopy, and DynCorp. The total amount the United States pays for private security is unknown.

Contracting is now part of the American way of war. Tellingly, Senator Obama sponsored a bill in 2007 to make armed contractors more accountable, and President Obama later ignored it. Hiring private military companies is one of the few issues in Washington that enjoys true bipartisan support, as Republican and Democrat White Houses use them more and more. Future American wars may be fully outsourced.

A BOOMING BUSINESS

The opening salvo of artillery was so intense that the American commandos dived into foxholes for protection. After the barrage, a column of tanks advanced on their positions, shooting their 125-millimeter turret guns. The commandos fired back, but it was not enough to stop the tanks.

A team of about thirty Delta Force soldiers and rangers from the Joint Special Operations Command—America's most elite task force—were pinned down at a Conoco gas plant in eastern Syria. Back at headquarters, roughly twenty miles away, a team of Green Berets and a platoon of Marines stared at their computer screens, watching the drone feeds of the battle. They were on a secret mission to defend the Conoco facility, alongside Kurdish and Arab forces. No one expected an enemy armored assault.

Attacking them were five hundred mercenaries, hired by Russia, who possessed artillery, armored personnel carriers, and T-72 tanks. These were not the cartoonish rabble depicted by Hollywood and Western pundits. This was the Wagner Group, a private military company based in Russia, whose employees, as with those at many high-end mercenaries, were organized and lethal.

The American commandos radioed for help. Warplanes arrived in waves, including Reaper drones, F-22 stealth fighter jets, F-15E strike fighters, B-52 bombers, AC-130 gunships, and AH-64 Apache helicopters. Scores of strikes pummeled the mercenaries, but they did not waver.

Four hours later, the mercenaries finally retreated. Four hours. No Americans were killed, and the US military touted this as a big win. But it wasn't. It took America's most elite troops and advanced aircraft four hours to defeat five hundred mercenaries. What happens when they have to face one thousand? Five thousand? More?

Mercenaries are more powerful than Westerners realize, a grave oversight. Those who assume they are cheap imitations of national armed forces invite disaster, because for-profit warriors constitute a wholly different genus and species of fighter. Private military companies like the Wagner Group or Blackwater are more like heavily armed multinational corporations than the Marine Corps. Their employees are recruited from various countries, and profitability is everything. Patriotism is unimportant, and sometimes a liability. Unsurprisingly, mercenaries do not fight conventionally.

Private force is big business, one that is global in scope. No one knows how many billions of dollars slosh around this illicit market. All we know is that business is booming. Since 2015, we have seen major activity by mercenaries in Yemen, Nigeria, Ukraine, Syria, and Iraq. Many of these for-profit warriors outclass local militaries, and a few can even stand up to America's most elite forces, as the battle in Syria shows.

The Middle East is swimming in mercenaries. Kurdistan is a haven for soldiers of fortune looking for work with the Kurdish militia, oil companies defending their oil fields, and those who just want terrorists dead. Some of these warriors for hire are just adventure seekers, while others are American veterans who've found civilian life meaningless. The capital of Kurdistan, Irbil, has become an unofficial marketplace of mercenary services, reminiscent of the Tatooine bar in the movie *Star Wars*—full of smugglers and guns for hire.

Private force has proved a useful option for wealthy Arab nations, particularly Saudi Arabia, Qatar, and the United Arab Emirates, all of which want to wage war but do not have an aggressive military. Mercenaries have fought on behalf of these countries in Yemen, Syria, and Libya in recent years. For example, the Emirates secretly dispatched hundreds of special forces mercenaries to fight the Iran-backed Houthis in Yemen. Hailing from Latin American countries like Colombia, Panama, El Salvador, and Chile, these men were all tough veterans of the drug wars who brought new tactics and toughness to the Middle East conflict. They were a bargain, too, costing a fraction of what an American or British mercenary would charge, so the Emirates hired 1,800 of them, paying each of them two to four times his old salary.

Turning the profit motive into a war strategy, Syria rewards mercenaries who seize territory from terrorists with oil and mining rights. At least two Russian companies have received contracts under this policy: Evro Polis and Stroytransgaz. These oil and mining firms then hired mercenaries to do the dirty work. For example, Evro Polis employed the Wagner Group to capture oil fields from ISIS in central Syria, which it did. Reports show there are about 2,500 Russia-bought mercenaries in Syria. Russia also uses them in Ukraine.

In fact, mercenaries are ubiquitous in the Ukraine conflict. The war there is awash in Russian, Chechen, French, Spanish, Swedish, and Serbian mercenaries, fighting for both sides in eastern Ukraine's bloody war. Companies like the Wagner Group conduct a wide range of secret missions, all of them denied by the Russian government. Ukrainian oligarchs have hired mercenaries, too, but not for their country's sake. For instance, the billionaire Igor Kolomoisky employed private warriors to capture the headquarters of the oil company UkrTransNafta, and one reason may have been to protect his financial assets.

In Africa, Nigeria secretly hired mercenaries to solve a big problem: Boko Haram. This Islamic terrorist group fought to carve out a caliphate in Nigeria, and the Nigerian army fought back, its methods no better. There is a saying in Africa: when elephants fight, the grass gets trampled. Tens of thousands of people were killed, and 2.3 million more were displaced from their homes. Boko Haram also abducted 276 schoolgirls for "wives," many of whom were never seen again. International outrage was swift but impotent.

Following the abductions, the Nigerian government secretly turned to mercenaries to fight Boko Haram. These were not the lone gunmen of B-grade movies, but a real private army. They arrived with special forces teams and Mi-24 Hind helicopter gunships—flying tanks. Conducting search-and-destroy missions, they drove out Boko Haram in a few weeks. The Nigerian military had not achieved this task in six years. Some wonder if the United States should hire mercenaries to hunt and kill terrorists in the Middle East, given the slow progress of national armies and the United Nations' absenteeism.

Even terrorists hire mercenaries. Malhama Tactical is based in Uzbekistan, and it works only for jihadi extremists. Malhama's hired guns are all Sunni, but not all are ideological, unlike their clients. Its services are standard for today's market, offering mercenaries that function as military trainers, arms dealers, and elite warriors. Most of their work is in Syria for the Nusra Front, an al-Qaeda-affiliated terrorist group, and the Turkistan Islamic Party, the Syrian branch of an Uighur extremist group based in China.

In the future, jihadis may hire mercenary special forces for precision terrorist attacks.

If terrorists can hire mercenaries, why not humanitarians? Nongovernmental organizations (NGOs) like CARE, Save the Children, CARITAS, and World Vision are increasingly turning to the private sector to protect their people, property, and interests in conflict zones. Large military companies like Aegis Defense Services and Triple Canopy advertise their services to NGOs, and NGO trade associations like the European Interagency Security Forum and InterAction provide their member organizations with guidelines for hiring them. Some think the United Nations should augment its thinning peacekeeping missions with certified private military companies. The option of private peacekeepers versus none at all—which is the condition in many parts of the world today—is a Hobson's choice. What's to stop a billionaire from buying a humanitarian intervention in the future? Ending a genocide would leave quite the legacy.

Multinational corporations, especially those in the extractive industries, are the biggest new clients of mercenaries. Companies working in dangerous places are tired of relying on corrupt or inept security forces provided to them by host governments, and they are increasingly turning instead to private force. For example, the mining giant Freeport-McMoRan employed the Triple Canopy firm to protect its vast mine in Papua, Indonesia, where there is an insurgency. The China National Petroleum Corporation contracts DeWe Security to safeguard its assets in the middle of South Sudan's civil war. Someday ExxonMobil or Google may hire an army, too.

There are mercenaries on the sea as well, similar to the privateers of two centuries ago. International shipping lines hire them to protect their ships traveling through pirate waters in the Gulf of Aden, the Strait of Malacca, and the Gulf of Guinea. Here's how it works: armed contractors sit on "arsenal ships" in pirate waters and chopper to a client's freighter or tanker when called. Once aboard, they act as "embarked security," hardening the ship with razor wire and protecting it with high-caliber firepower. After the ship passes through pirate waters, the team returns to its arsenal ship and awaits the next client. The industry is based in London, and it seeks legitimacy through ISO 28007 certification. Some would like to see the return of true privateers: private naval vessels that sail under letters of marque, issued by a government, who are authorized to attack and plunder enemy ships, such as pirates. Americans will be pleased to know that Congress is authorized to hire privateers under article 1, section 8, of the Constitution, and this could prove more efficient than sending navy destroyers after pirate Zodiacs.

There are even mercenaries in cyberspace, called "hack back" companies. These computer companies attack hackers, or "hack back," those who assail their client's networks. Hack back companies cannot undo the damage of a network breach, but that's not the point. They serve as a deterrent. If hackers are choosing targets, and they know that one company has a hack back company behind it and another does not, they select the softer target. Also known as active defense, this practice is currently illegal in many countries, including the United States, but some are questioning this wisdom, since the National Security Agency offers scant protection for nongovernment entities. For example, the WannaCry ransomware attack in 2017 infected more than 230,000 computers in over 150 countries. Victims included the United Kingdom's National Health Service, Spain's Telefónica, Germany's Deutsche Bahn, and US companies like Federal Express. If countries cannot protect their people and organizations from cyberattacks, then why not allow them to protect themselves?

Private force is manifesting itself everywhere. After one hundred-fifty years underground, the market for force has returned in just two decades, and it is growing at an alarming rate. In military strategy, there are five domains of war: land, sea, air, space, and cyber. In less than twenty years, private force has proliferated among every domain except space, but that, too, may change. Space exploration is already privatized, for example, with companies like SpaceX, and it is possible that private armed satellites may one day orbit the earth.

Worse things are to come. In just ten years, the market for force has moved beyond Blackwater in Iraq and become more lethal. Mercenaries are appearing everywhere, and they're no longer just on the fringe. Contract warfare has become a new way of warfare, one resurrected by the United States and imitated by others.

THE FUTILITY OF LAW

Recently I was in London, giving a talk about the private military industry. The eighteenth-century room was magnificent, the audience a mixture of defense intellectuals, graduate students, and government officials. The questions were polite but sharp, in a British sort of way. Too often, regardless of where I am, I get a question like this. Actually, it's more of a statement:

"But it's illegal!"

"Saying something is illegal isn't a strategy," I say. "It's an appeal to authority, and whose authority? The UN has abdicated its role of conflict prevention, its main mission since 1946. It has done nothing to stop the growth of mercenaries, just like it did nothing meaningful to stop the wars in Iraq, Syria, Sudan . . . everywhere. International law does not work, nor does the UN. Both will become shells that others will ignore or twist for their own purposes."

The room sat still, like I had just called the queen a Drury Lane strumpet.

International public law is fiction. One famed legal scholar called it the "vanishing point of law" while others ridicule it as "soft law" because it is followed by courtesy rather than necessity, unlike real laws. Why? Because international law is 60 percent diplomatic custom and 40 percent nonbinding treaties between states—it's just diplomacy. There is no international judiciary, police force, or prisons, so it really doesn't matter if you ignore it.

A hand goes up.

"But if the world really wanted to stop mercenaries, it could. It would forcibly outlaw companies like Blackwater and hunt down the rest."

Heads nod across the audience.

"Maybe. But who is going to go into a war zone and arrest all those mercenaries? The UN? NATO? Is the US really going to deploy the 82nd Airborne Division into Yemen or Nigeria to arrest mercenaries? I don't think so."

Heads stop nodding.

"Besides," I add, "mercenaries will shoot law enforcement dead. You can't regulate something that chases off law enforcement."

Another hand shoots up, furiously waving at me.

"I'm a lawyer," the man says.

Oh crap, I think.

"The easiest way to stop mercenaries is to go after their clients. Legally go after them, I mean. No one will hire mercenaries if it's a one-way ticket to jail, and that will solve the problem."

Heads start nodding again.

"Don't count on it," I say. "Many buyers are states like Russia, Nigeria, the Emirates, and the US. It will be difficult to send them to private war jail."

"But it's the nonstate actors I'm talking about," the lawyer says, almost standing up.

"They're even trickier. You push them too hard and they move offshore, beyond the reach of the law. Big corporations already do this to evade taxes."

Heads stop nodding again. Another hand goes up.

"So, what do you think we should do?"

It's my least favorite question, because it's the hardest.

Some have argued for self-regulation, but this is asking the proverbial fox to guard the henhouse. An organization based in Geneva called the International Code of Conduct Association already exists to do this. To join, potential buyers and force providers (they avoid the *M* word) swear an oath to obey laws and respect human rights. It's a noble model, but the market for force is passing it by.

Others have suggested market solutions. Superbuyers can use their market power to shape the industry's behavior by rewarding good force providers with lucrative contracts while driving the rest out of business. Who is a superbuyer? The United States was during the Iraq and Afghanistan Wars, and the United Nations could be, if it privatizes some of its peacekeeping missions. Alternatively, a cartel of buyers could become a superbuyer. However, cartels are tough to maintain, because defection is cheap while holding fast is not. Or we could do what states did in 1648 and monopolize the market. For this to work, all countries would have to pool their resources to abolish mercenaries. World peace might be easier.

"Frankly," I say, "we need to accept the fact that mercenaries are here to stay, and they will change warfare as we know it."

RULE 7: NEW TYPES OF WORLD POWERS WILL RULE

Megachurches are a god-force of faith. These super-size Protestant houses of worship are found across the United States, especially in the South. Their preachers are like rock stars, sermonizing on multiple jumbo screens inside churches the size of sports stadiums. One megachurch in Houston, Texas, boasts a congregation averaging about 52,000 attendees per week. Sermons are televised, reaching another seven million viewers weekly and twenty million monthly in over one hundred countries.

American megachurches lag behind the largest congregations around the world. The Bethany Church of God in Indonesia attracts 140,000 worshipers a week, Calvary Temple Church in India has 130,000 believers, and Nigeria has four megachurches in Lagos alone that average 50,000 believers each. The Yoido Full Gospel Church in Seoul, South Korea, claims a whopping 480,000 members.

Megachurches command big bucks. "If you put together all the megachurches in the United States, that's easily several billion dollars," says Scott Thumma, a professor of sociology and religion at Hartford Seminary. He estimates that the average American megachurch "has about $6.5 million in income a year." The Lakewood Church in Houston has an annual budget of $90 million.

Here's a hypothetical situation: Muslim terrorists in the Middle East are crucifying Christian men, selling Christian women and girls into sexual slavery, and blowing up Christian churches. People around the world are horrified, yet the international community does little to stop it. Many Christians believe they have a duty to defend the defenseless, especially if it means protecting fellow Christians and holy sites. This mandate helped launch the Crusades nine hundred years ago, after caliph Al-Hakim destroyed churches and religious artifacts in Jerusalem, including the Church of the Holy Sepulcher.

What if a megachurch sponsored a crusade today? Evangelicals are not pacifists, and some are even militant in their defense of their faith. They could easily hire mercenaries to protect endangered Christian communities. Some might even employ a small private army to destroy terrorists.

Concerns over legality are secondary. Evangelicals believe the laws of men are eclipsed by duty to God, and many consider Muslim terrorists like ISIS unholy abominations worthy of obliteration. Additionally, the international community's failure to protect the innocent could be seen by such individuals as a moral abdication justifying global vigilantism. Christians have a higher calling. It just takes one charismatic preacher in a megachurch to make this argument, and—voilà—a crusade is born.

This scenario is not purely hypothetical. A grassroots Christian mobilization occurred during ISIS's takeover of northern Iraq. The city of Irbil, in the autonomous region of Kurdistan, became a hub for foreign fighters looking to kill ISIS. Some were mercenaries, while others were Christian crusaders. Many were American and British veterans of the Iraq War.

The Christian resistance grew after the United States pulled out of Iraq. Knowing what lay ahead, militia units like the Nineveh Plain Protection Units, the Tiger Guards, the Babylon Brigade, and the Syriac Military Council took shape. Additionally, churches from around the world sent missionaries to assist beleaguered communities on the front lines, supported by their congregations back home.

One such example of a Christian militia is Dwekh Nawsha, whose name means "one who sacrifices" in Assyrian. It has a foreign-fighters battalion, and one of its recruiters is a twenty-eight-year-old who goes by the pseudonym "Brett." Detroit born and bred, he is a US Army veteran who fought in the Iraq War. Now he is a self-described "soldier of Christ" and a "crusader." On his left arm is a tattoo of a machine gun, and on his right, one of Jesus in a crown of thorns.

"Jesus says, you know, 'What you do unto the least of them, you do unto me,'" Brett said. "I take that very seriously."

Imagine what would happen if a megachurch sponsored a grassroots crusade like this? That is, what if such a church hired high-end mercenaries to destroy ISIS's core? They would go to war against ISIS, or whatever follows it, and maybe even win. Like mercenaries, crusaders may no longer be relics of the past.

If this future sounds crazy to you, know that it's closer than you realize. Various billionaires and humanitarian organizations have already tried hiring mercenaries to stop wars on their own. A few years ago, I was asked to participate in such a plan. Mia Farrow, the millionaire actress, approached Blackwater and a few human rights organizations to end the genocide in Darfur, Sudan. The plan was simple. Blackwater would stage an armed intervention in Darfur and establish so-called islands of humanity, refugee camps protected by mercenary firepower. These would be safe havens for refugees fleeing the deadly *janjaweed*, gunmen who massacre whole villages in Darfur. During this time, the human rights organizations would mount a global name-and-shame media campaign to goad the international community into ending the genocide once and for all with a muscular UN peacekeeping mission. Ultimately, Farrow chose not to go through with the plan, but that's not the point. The plan was doable, and anyone with money could pull it off.

That was in 2008, when many were still gun-shy about hiring mercenaries. No longer. Mercenaries are back, and hiring them is becoming more common. Yesterday it was Mia Farrow; who will it be tomorrow? Any millionaire or organization looking to leave a legacy can write a check and end a war. Or start one. Or place bounties on terrorists' heads. However, their actions could also suck the United States and others into unwanted quagmires. Unintended consequences are a constant of war.

One can begin to see a medieval universe unfolding, in which nations, churches, and the wealthy each pursue global ambitions as world powers. They will all use force when necessary because it can be bought once again, as in the Middle Ages. The use of private force will expand in the decades to come, because nothing is in place to stop its growth, and in so doing, it will turn the super-rich into potential superpowers.

THE RETREAT OF STATES

One of my jobs in Liberia was putting warlords out of business. They were ferocious, and all had cute noms de guerre: General Mosquito, General Peanut Butter, Superman. My favorite was General Butt Naked, whose men fought au naturel. There was a cross-dressing militia, too, who donned feather boas and toted dainty purses into battle. After the war, I ran into Butt Naked on the streets of Monrovia, Liberia's capital, preaching the gospel.

Women also make formidable warlords, despite the prognostications of feminist theory. A twenty-year-old woman named Patricia was gang-raped by paramilitaries, then left for dead. She resurrected herself as "Black Diamond" and formed an all-women militia. Armed with AK-47s and RPGs, they unleashed holy hell on would-be rapists and took what they wanted.

"If you are angry," Black Diamond told me, "you get brave. You can become a master in everything." Later, I heard she had a love child, "Small Diamond."

The warlord in chief of Liberia was Charles Taylor, who also happened to be its president. Like all warlords, Taylor rose to power by seizing it. But Taylor was smarter than most, and he realized that if he was elected, the international community would accept him as the legitimate ruler of Liberia, rather than just a thug.

So, after Taylor took the executive mansion by force, he held elections, rolling out the catchy campaign slogan "He killed my ma, he killed my pa, but I will vote for him." And people did. Taylor won by a landslide, with 75 percent of the vote. The United Nations and the Carter Center declared the election free and fair.

"Why did you vote for Taylor?" I asked one Liberian.

"Because if we didn't, he would have killed us all," said the man, summing up the fear of a nation held hostage.

President Taylor, now with a vote in the United Nations, entered formal relations with other countries and enjoyed legitimacy. But it was all perversion. Taylor, like the warlords before him, saw the country as a prize to be pillaged. And pillage he did. He looted $100 million from the treasury and lived lavishly as Liberians starved. He had a small army of child soldiers who revered him as a father-god, and he traded so many "blood diamonds" (gems harvested in the gore of war), that the phrase became a household term around the world. Taylor also liked war bling, and he had a chromed AK-47, which I held years later. If he saw a woman he wanted, he would send men to take her. After he was done, she may or may not have lived. His son Chucky was worse. Men who stood up to the Taylors disappeared or were eaten: many warlords practiced ritual cannibalism.

Charles and Chucky Taylor are no more, ousted by rebels at gunpoint, but Liberia remains a fragile state on life support. Like that of all fragile states, Liberia's government is not wholly in control. Yes, the country has a president, a presidential palace, a flag, and a seat at the United Nations. But it lives off international aid, its social services are administered by foreign NGOs, and it's essentially occupied by a large UN peacekeeping force. The government's authority does not fully extend to the country's borders, which creates pockets of ungoverned space where anything can fester—and things do. Warlords wait in the wings, and the army waits for them (something I had a hand in). Corruption is institutionalized. Liberia is not a state, but the charade of one.

Many people think failed states are the exception in world affairs, but they're the rule. Most of the world's 194 states are fragile, like Liberia. Some are better off, others worse. Some have failed outright, like Syria and Somalia. Diplomats don't like to discuss this topic at cocktail parties, but it's a BFO—a blinding flash of the obvious—for those who travel where *Fodor's* dares not go. Dozens of studies confirm this. Billions of people live in countries that are in danger of collapse, but that doesn't mean anarchy. As one kind of authority withdraws, another will advance.

State erosion encourages new kinds of global powers. The vacuum of authority left by retreating states will be filled by insurgents, caliphates, corporatocracies, narco-states, warlord kingdoms, mercenary overlords, and wastelands. For example, after Israel left southern Lebanon in 2000, it wasn't the government of Lebanon who took over the land but Hezbollah, a terrorist group. When Israel invaded southern Lebanon in 2006, it fought Hezbollah, not the Lebanese army. Southern Lebanon is still owned by Hezbollah. This is durable disorder in action.

There are other examples. In the wake of northern Iraq's disintegration, the Kurds established a de facto independent Kurdistan, defended by their own army, the Peshmerga. Somalia has balkanized into Somaliland, Puntland, and what's left of the rest, all vying for power with their own paramilitaries. South American drug lords have taken over the West African country of Guinea-Bissau, making it a narco-state and a hub for drugs headed to Europe. Whole swaths of Africa are ruled by unknowns, or by no one at all. The illusion of states may continue on maps but not in reality, as new types of powers slowly take over.

The international community's response to failing states—nation-building—has been disastrous. The United States' efforts in Iraq and Afghanistan are good examples. Iraq was better governed under Saddam Hussein, and Americans have learned why Afghanistan is called "the graveyard of empires." The United Nations' record is equally dismal. Its peacebuilding missions in South Sudan, the Democratic Republic of the Congo, Burundi, Haiti, and other places have been costly and hopeless. None of those states are stronger because of UN help. Nation-building fails because countries are not machines that can be built.

As with mercenaries, the retreat of states is both a symptom and a cause of durable disorder. States will not disappear, and the top twenty-five countries such as the United States and Western Europe will remain strong. The bottom twenty-five like Myanmar and Haiti will go the way of Somalia. What of the middle? In the future, countries will look more like counties. They will be in charge of roads and bridges, but little else. People will come and go as they please, and they will do what they want. The era of absolute state rule and the Westphalian Order climaxed in the twentieth century.

For those who lament the passing of the state, weak states have at least one option. They can hire mercenaries to reconquer their land and reestablish their rule of law, as Nigeria did against Boko Haram. Expect to see more of this in the future. However, many states are content to remain corrupt regimes, preying on their people rather than serving them, all the while living off international charity. The retreat of states will embolden new kinds of superpowers to fill the vacuum of authority.

MEET THE NEW ELITE

Mia Farrow is a member of a growing club. We are entering a reality in which anyone with enough money can rent a military to do whatever he or she wants: start wars, end them, seize other people's property, murder whole groups of people or save them. What makes this possible is the enabling power of mercenaries. When anyone can hire mercenaries to wage war, the ultrarich can become a new kind of power in world affairs.

This may sound preposterous, but it's not. Actually, it's the law of supply and demand. War is becoming marketized, and mercenary supply will attract new demand through the invisible hand of the marketplace. But that demand won't just be states; it will be anyone who has money and wants security. That's a lot of people. We live in an increasingly insecure world, and mercenaries sell security. Clients will emerge, and some will grow powerful in the decades to come. They will become a new ruling class in international relations.

Who will be in this new class of world powers? The global 1 percent, for starters. In 2015, just sixty-two individuals possessed more wealth than the poorest half of the planet. In other words, a busload of tycoons owns more than 3.6 billion people, combined, making these magnates the 0.000002 percent. The consolidation of apex wealth is a growing trend. In 2010, there were three hundred eighty-eight billionaires whose wealth was equal to the bottom half of humanity; now it's sixty-two. In 2017, Credit Suisse revealed that the richest 1 percent have now accumulated more wealth than the rest of the world put together.

Private force allows *Forbes*'s list of billionaires and the Fortune 500 to become armed and dangerous. Already they are more powerful than most states. According to the World Bank, the top one hundred economies comprise thirty-one countries and sixty-nine corporations. Walmart has the world's tenth largest economy, ahead of India (twenty-four) and Russia (thirty). Can anyone really argue that Gabon is more influential in world affairs than ExxonMobil simply because it's a state? Of course not. Now ExxonMobil can have its own army, too, making it even more powerful.

From an oil company's perspective, hiring mercenaries makes good business sense. For decades, companies like ExxonMobil and Shell have been tethered to corrupt governments such as Nigeria's for their security, risking the lives of their employees and losing money. In 2013, terrorists attacked a natural gas facility in Algeria operated by BP and Statoil, killing forty employees. This would not have happened if the oil companies had mercenaries guarding the site.

In the future, megacorporations and the global 1 percent will invest in their own security, given that mercenaries will be available and legitimized. Pressure from shareholders alone will dictate this. Some may argue that this is illegal, but international law is feeble, and the political will to enforce it is weak. Who is going to arrest the leadership of a rich oil company and capture its mercenaries in a place like sub-Saharan Africa or the Middle East? No one. Some say such a scenario is unlikely, but a Chinese oil company is already hiring mercenaries in South Sudan, and so is a Russian mining company in Syria, as discussed above. Others will follow.

The number of countries hiring mercenaries will also increase.

Every year more and more governments hire them. First it was the United States, and then Russia. Now countries in the Middle East and Africa are turning to private force for their security needs. This trend has de facto legitimized private force. It's only a matter of time before everyone hires mercenaries, as was done just a few centuries ago. Even the UN might employ private force to augment its anemic peacekeeping missions. Soon, all will rent private armies, commodifying conflict worldwide. This condition will only breed more mercenaries, as supply rushes to meet demand.

Organized criminals can become superpowers, too. Oligarchs and drug cartels already rely on militias and gangs for muscle, but now they can rent industrial-strength firepower, such as attack helicopters and private military regiments, to capture states and turn them into puppets. This will accelerate what is already happening around the world. Narco-states exist throughout Latin America and West Africa, and much of the Eurasian countries are mafia states. Mercenaries will make drug wars and oligarchs' conflicts far bloodier. In Mexico, 11,155 people were killed in drug-related violence during the first five months of 2017, about one every twenty minutes. How many more would have died if the competing cartels had mercenary firepower? Mercenaries enable criminal networks to replace states, not just hide behind them.

Terrorism can also get worse, with mercenary help. A niche black market may be developing that offers coveted services like training and equipping terrorist forces, conducting strategic reconnaissance of potential targets, and performing direct actions or terrorist attacks. As mentioned earlier, a mercenary group called Malhama Tactical based in Uzbekistan already does this. This is frightening on many levels. To date, most terrorist attacks have been the work of zealous amateurs. A mercenary special forces team would be far deadlier, capable of creating a mass-casualty event that would be difficult to detect and defeat. Also, the fall of ISIS taught terrorists that if they take land, they must be prepared to hold it. The forces of jihad were too weak to do so, but mercenaries would rebalance the conflict. Worse, the skills of quality mercenaries are a force multiplier. Groups like al-Shabaab in the Horn of Africa, Boko Haram in West Africa, and al-Qaeda worldwide may seek professional help in their jihad against the West. Horrible as it sounds, mercenaries could launch terrorists into a new league of awful.

However, people in the future could fight back with mercenaries of their own, as Mia Farrow considered doing. Megachurches, large NGOs, well-funded mosques, and concerned billionaires could hire mercenaries for messianic do-good missions. Mercenaries are eager to help, too, as many of them would rather serve angels than devils, as long as the pay is good. Private force can stage armed humanitarian interventions to stop genocide, launch rescue missions to save refugees, safeguard humanitarian convoys from ambush, protect charities working in conflict zones, or assassinate leaders who incite mass killings and other gross human rights violations. How many atrocities could have been mitigated in this way? Perhaps Rwanda, the Balkans, Darfur, and the Yazidis. Certainly, doing nothing proved ineffective. Causing some harm to enact great good is a moral calculus made possible by mercenaries, and potential clients will attempt to do so as the international community falters in the face of genocide.

When the super-rich become superpowers, what will it look like? The British East India Company may prove instructive. For all the power wielded today by the world's largest corporations—

whether ExxonMobil, Walmart, or Google—they are tame beasts compared with the British East India Company. Founded in 1600 as a joint stock company, it became the greatest corporation in history and the original corporate raider. One of the very first Indian words to enter the English language was the Hindustani slang for plunder: "loot." In its 275-year run, the company conquered India for the British crown, although at times it was hard to distinguish who served whom.

What made the British East India Company powerful was its private military. With its own armed forces, it conducted the military conquest, subjugation, plunder, and rule of a subcontinent, while fending off its European rivals. By the turn of the nineteenth century, the company boasted a mercenary army of 150,000 soldiers and 122 ships of the line, the larger ones mounting up to forty guns, a match for all but the most powerful enemy warships. However, in a warning to tomorrow's aspiring superpowers, maintaining this private military bankrupted the company.

The world will change when any organization can hire an army to do its bidding. International relations as we know it—state-on-state engagement—will grow obsolete, because private force will enable new superpowers that are not states. But what happens when private force becomes its own superpower?

MERCENARY OVERLORDS

Mercenaries can also go into business for themselves. Why take orders when you can give them? As with narco-states and mafia states, strong mercenaries can coopt and even take over whole countries, turning them into "merc-states." Mercenary overlords are not new. In the Middle Ages and Renaissance, mercenary captains often installed themselves as rulers, as happened in Perugia, Rimini, Urbino, and Camerino. One cunning mercenary nicknamed Sforza ("Force") took over Milan and founded a dynasty that ruled it for almost a century.

Regions ripe for mercenary takeover are those rich in natural resources but weak in government. Such places are everywhere: the oil fields of Venezuela, the mines of the Congo, the timber forests of Liberia, the natural gas reserves of Yemen, and the sweet crude of Libya. Wherever there is war, natural resources, and someone willing to buy those resources, expect the potential for a merc-state. The world's easy areas have all been mined or drilled, leaving only conflict regions untapped. This represents an opportunity for enterprising mercenaries. Some think the development of mercenary governments is impossible because it's illegal, and they assume that the international community would rush to stop it. Don't be so sure. The world is already moving in this direction, as evidenced by the rise of mercenaries and clients. A merc-state simply cuts out the middlemen.

How would mercenaries take over a state? Many ways. For starters, there's the old-fashioned method of conquest. A well-armed mercenary cadre could carve out a piece of Somalia or Yemen, both of which have untapped oil and gas reserves owing to the durable disorder there. Alternatively, mercenaries could stage a coup d'état. Some coups are purely internal affairs, but many have outside help. Since 1950, there have been over two hundred successful coups, and thirteen world leaders in 2018 had seized control via firepower. Once installed, such leaders often declare themselves president for life and rule like feudal kings. Mobutu Sese Seko, the usurper of the Congo, even changed his name to mean "Rooster Who Gets All the Hens" in the local language. Generals and warlords take over countries every day and get away with it, so why not mercenaries?

Alternatively, mercenaries could hijack a separatist movement in a resource-rich area, such as Katanga Province in the Democratic Republic of the Congo or Aceh in Indonesia. In this scenario, one overlord replaces another. The mercenaries initially side with the rebels, then turn on them after the government is expunged from the region. Once the mercenaries control the asset in question—uranium, oil, gas, copper, gold, cobalt, lithium, timber, diamonds—they then sell it on the black market or use shell companies to move it on the open market. The locals would serve as the workforce, little more than slaves.

A clever mercenary could also stage a palace coup and rule from behind the throne (or executive armchair). This option saves the hassle of an invasion by holding a leader hostage in his own capital, while subduing or bribing a country's existing armed forces. This is called "praetorianism," a name that comes from the infamous Praetorian Guard, the imperial bodyguard of the Roman emperors established by Augustus Caesar. But rather than protect the emperor, the guard often controlled him. During its three-hundred-year existence, the Praetorian Guard assassinated fourteen emperors, appointed five, and even sold the office to the highest bidder on one occasion. Why deal with the headache of running a state when you can make it someone else's problem? Plus, you get a vote in the United Nations and look more legitimate in the eyes of other world leaders.

Mercenaries can exploit durable disorder for profit, not just on the battlefield but also as rulers. This scenario breeds more chaos and shows how durable disorder feeds on itself.

DEEP STATES

There is another kind of new superpower, and it's not really new—but its role will change as states recede. Do you ever wonder why some government policies remain the same no matter who the president or prime minister is? Especially when that person campaigned against those same policies as a candidate? And those policies are unpopular? For example, Presidents Obama and Trump both campaigned against the war in Afghanistan, only to expand it shortly after taking office. If this mystifies you, you are not alone. People around the world are perplexed at their government's unremitting embrace of damaging policies. There may be a deep state at work.

Deep states exist, but not as you know them. They are not conspiracies or products of conspiracy theory. Conspiracies seek to overturn the establishment, while the deep state *is* the establishment, and then some. Deep states are like states with cancer. Their institutions of power—the military, the judiciary, intelligence agencies—have gone rogue; rather than serving the state, they make the state serve them. Over time, they end up usurping power in an internal coup d'état, becoming a "state within a state" that influences policy without regard for legitimate leadership or the concerns of citizens. In other words, these renegade institutions hijack the nation from deep within the state's own structure.

When I teach this topic at Georgetown University, my graduate students initially reject the idea of a deep state as a conspiracy theory. To be fair, that's often how it's portrayed. Frustrated politicians and activists blame the deep state for undermining them, and people roll their eyes. The concept of the deep state has been around for years, but the term was unknown to most in the West and especially the United States until recently. President Trump's alt-right defenders and his former chief strategist Steve Bannon have blamed the deep state for trying to delegitimize the president. Articles in *Breitbart News*, where Bannon served as executive chairman before and after working for Trump, have invoked the idea repeatedly. Critics balk at this accusation, reaching for tinfoil hats while lampooning alt-righters as paranoid weirdos. Both sides are wrong—deep states are real, but they are not conspiracies.

The distinction between a conspiracy and a deep state is subtle yet profound. It's really about individual versus institutional actors. Conspiracies are powered by individuals, known as conspirators. Orchestrated by a cabal or a mastermind, conspirators pool their personal connections, influence, money, and other resources to subvert the establishment. This is why conspiracies must hide in the shadows, for self-preservation. Should the establishment catch them, they would be branded traitors and hung.

Great conspiracies of the past include Guy Fawkes and the Gunpowder Plot to blow up England's king and parliament in 1605, and John Wilkes Booth's assassination of Abraham Lincoln after the American Civil War, meant to revive the Confederate cause. In both cases, the conspirators were hunted, caught, and killed, their deaths serving as a warning to other would-be threats to the establishment. It's worth noting that past conspiracies still inspire. Today, antiestablishment protesters ranging from members of the hacktivist group Anonymous to Occupy Wall Street demonstrators don Guy Fawkes masks as a symbol of dissent. The mask has become iconic in modern political culture, popularized by the comic book series and movie *V for Vendetta*, which features a Fawkes-masked vigilante named V who battles a fictional fascist English state. Vigilantes like V are models of antiestablishment heroes, and they constitute examples of how conspiracies fight the system.

Unlike conspiracies, deep states are institutional actors. Yes, institutions are populated by people, but they are not the same as conspirators. There is an old joke about bacon and eggs: the chicken was involved but the pig was committed. Conspirators are like the pig, risking everything. People in institutions are like the chicken, who toil for the cause but rarely make the full and final sacrifice. Conspirators risk their personal resources, whereas institutional employees deploy an organization's assets. A general doesn't fight the enemy using his private wealth; he uses the army. If he is defeated, the general gets fired, but the army marches on. Fawkes and his coconspirators lost and became bacon.

In some ways, institutions are their own organism. They don't let just anyone rise to the top. Only individuals who have been institutionalized after decades of service are promoted to the highest ranks, where they will then reliably promote the institution's agenda. Conspirators would deride these "company men" as "empty suits" who are promoted for their groupthink—institution *über alles*!—that is, until they were caught and hung by those same company men.

What are deep state institutions? It's different in every country, and not all countries have them. But generally, they are the institutions of power: the military, the secret police, the intelligence services, the law enforcement agencies, and the judiciary. Their authority is codified in law, making them completely legal. What makes them different from normal institutions is that they've gone rogue. Deep state institutions place their own interests above that of the state and its citizens. There may be a legitimate government in place, but it's the deep state that really calls the shots.

The institutions that comprise the deep state do not plot their actions like participants in a conspiracy. Rather, they engage in passive synchronization. They cooperate because their institutional interests align, as in a Nash equilibrium, resulting in mutually reinforcing actions that protect their common goals. Gradually this tacit consensus congeals into a deep state that can control a nation. They can overrule, sabotage, and reverse legitimate government decisions with no accountability or even visibility.

Conspiracies and deep states are natural enemies. Conspiracies seek to undermine the system, while deep states seek to hijack it. Conspiracies hide in the shadows, while deep states operate in the open. Conspiracies are composed of radical individuals, while deep states are institutions. The time frame for conspiracies is short, usually months or years. Deep states think in terms of decades and centuries. In social science parlance, conspiracies represent agency while deep states embody structure, in a demonstration of the classic structure-versus-agency debate. Conspiracies and deep states are as different as fire and earth.

Deep states are not new. In the early Middle Ages, the Frankish Merovingian dynasty was governed not by its kings, but by the "mayors of the palace," or majordomos. Other examples include the peshwa in India's Maratha Empire, the shogun in feudal Japan, and the prime ministers of England in the eighteenth and nineteenth centuries. These were the true loci of power behind the official rulers.

The term "deep state" comes from Turkey, and it became popular in the 1990s. It has since become central to Turkish scholars, citizens, and observers to describe how that country's government truly works. Even US diplomats use the term to explain Turkish politics to Washington. There are many accounts of the Turkish deep state (*derin devlet*), but most describe it as the behind-the-scenes machinery and power relationships among institutional elites in the military, the intelligence services, the law enforcement agencies, the judiciary, and the mafia. Together, these organizations have instituted a permanent national security apparatus in Turkey that privileges them with everlasting "emergency" authority that undermines that of the elected government.

The Turkish deep state is not an idiosyncrasy but a model. Political scientists and foreign policy experts use the deep state to describe institutions that exercise power independent of, and sometimes over, legitimate political leaders. Sometimes the concept of the deep state is the only way to rationalize the behavior exhibited by authoritarian countries like Turkey, Algeria, Pakistan, Egypt, and Russia, where generals and spymasters are the true rulers in nominally democratic societies, even replacing elected leaders when they see fit. Saddam Hussein's Iraq was a Ba'athist deep state, and the Arab Spring was a popular revolt against deep states across North Africa and the Middle East. The deep states either crushed or bought off protestors demanding democracy. None fell. In 2016, the governments of Turkey and Egypt moved to more overt security-state dictatorships, in which the deep state is the only state.

Iran is a classic deep state, a theocracy with a fig leaf of representative government. "It has essentially two states," according to David Petraeus, who commanded US forces in the region for years. There is the visible state of an elected president, a parliament, ministers, an army, a navy, an air force, and marines. Then there is the deep state composed of the Revolutionary Guards Corps–Quds Force, with its army, navy, air force, and marines, and its overseas adventures. "These two sides are in a fair amount of tension with one another," says Petraeus.

Western liberal democracy may also have its share of deep states. This idea is quite old, just forgotten. In the 1860s, Walter Bagehot unveiled his theory of "double government" in his book *The English Constitution*. No political slouch, he was the editor in chief of the *Economist* magazine for seventeen years, and the magazine still has a column named in his honor. Bagehot suggested there are two sets of institutions, which together form a double government. Dignified ones "impress the many" and efficient ones "govern the many." The dignified institutions were the monarchy and the House of Lords, which people erroneously believed ran the government. In reality, the true seat of power lay with the efficient institutions: the House of Commons, the prime minister, and the cabinet. Back then, they labored in the shadow of Queen Victoria and her nobles, who claimed to rule all.

What is formally written in the constitution is irrelevant, insists Bagehot. All that matters, he argues, are the efficient institutions, an idea echoing that of the modern deep state. Things have changed quite a bit since the 1860s. Today, no one would mistake 10 Downing Street for a deep state. The monarchy has slipped to the gossip pages, and the House of Lords is a public relic. Meanwhile, the House of Commons, the cabinet, and prime minister have evolved into the new dignified institutions. What are the efficient institutions, or the deep state? Some would say it's the British civil service.

Lovers of British comedy will remember the BBC series *Yes, Minister* from the early 1980s. It features a cabinet minister, Jim Hacker, who is continually outfoxed by the British civil service, especially his permanent secretary, Sir Humphrey Appleby. Hacker is Faust to Appleby's Mephistopheles. Hacker is an affable fellow who genuinely wishes to do public good but cannot get anything done, as he's always checkmated by bureaucrats. The well-connected Appleby seems nice enough but always devises some clever machination to hijack the minister's political agenda. He believes that the civil service, being politically impartial and experienced in government, knows what is best for the country—a belief shared by his bureaucratic colleagues. Fed up, in one episode Hacker seeks out the advice of his predecessor, a member of the opposition party, Tom Sargent:

> **JIM HACKER:** Look, Tom. You were in office for years. You know all the civil service tricks.
>
> **TOM SARGENT:** Oh no, not all, old boy. Just a few hundred.
>
> **JIM HACKER:** How do you defeat them? How do you make them do something they don't want to do?
>
> **TOM SARGENT:** My dear fellow. If I knew that I wouldn't be in opposition.

Like all satire, it is funny because it's half-true. Margaret Thatcher, then the real prime minister, was a fan of the show, and she said its "clearly-observed portrayal of what goes on in the corridors of power has given me hours of pure joy." She even performed a *Yes, Minister* sketch. The series remains a classic of political satire and received numerous awards, including several BAFTAs.

It is possible that the United States has a deep state. Free-market democracy has been central to the republic since its founding, entwining political and business interests. Examples are numerous, from the undue political weight of the Vanderbilts, the Rockefellers, and other magnates during the Gilded Age, to today's megacorporations, whose heavy lobbying activities receive the same legal treatment as a person exercising free speech. The American political system has always protected the nexus of corporate and political interests.

Concern that Wall Street trumps Main Street in America's democracy is an old one. Tycoons and members of Congress dismiss it all as conspiratorial claptrap. However, the unaccountable power of big business and their dark money sloshing around Capitol Hill has long provoked citizen ire, from labor protests in the 1870s to the recent Occupy Wall Street protests. In the presidential election of 1896, the robber barons of the age—John D. Rockefeller, Andrew Carnegie, J. P. Morgan—bankrolled the Republican nominee, William McKinley, to defeat Democratic nominee William Jennings Bryan, and threatened their workers with lost jobs and closed down factories if Bryan won. McKinley was elected, but was assassinated five years later. The plutocrats met their match with the new president, Theodore Roosevelt, who was the first White House occupant to say it aloud: "Behind the ostensible government sits enthroned an invisible government, owing no allegiance and acknowledging no responsibility to the people."

The secret marriage of corporate and political agendas birthing an American deep state may shock some, but it shouldn't. President Eisenhower famously warned the nation against the corrupting influence of what he termed the "military-industrial complex" in his farewell address. The military-industrial complex is a deep state alliance among the military, the arms industry that supplies it, and Congress, which oversees it. It's an infinity loop fueled by corporate contributions to politicians, congressional approval for military spending, lobbying to support bureaucracies, and pliant government oversight of the industry.

Although the conflicts of interest are in plain sight, it does not stop retired generals and admirals from sitting on boards of "Beltway bandits": corporations that line Washington's ring highway and sell equipment to the Pentagon. These individuals help ease along lucrative military contracts. The result are more F-35s and aircraft carriers—the most expensive yet superfluous weapons in history—and the spending buffet of the Third Offset strategy. Ultimately, the deep state of the military-industrial complex encourages the militarization of foreign policy, forming a challenge to world peace.

A favorite tactic of deep state doubters is to deride the idea as fringe kooky, but Eisenhower's credibility is beyond reproach. He was a two-term president, a retired five-star general, and a hero of World War II—he had unparalleled authority in the matter. The language he used to describe the military-industrial complex mirrors how we think about deep states today. "We must guard against the acquisition of unwarranted influence . . . by the military-industrial complex," he said, adding, "The potential for the disastrous rise of misplaced power exists and will persist." For his sins, the military, with congressional approval, renamed its top war college for arms procurement after him. At least the deep state has a sense of humor.

There have been other warnings about the American deep state since then. Mike Lofgren was a congressional aide for twenty-eight years who retired in 2011 after serving on both the House and Senate budget committees. Those are some of the most powerful committees in Congress, because they oversee the government's $3.8 trillion budget, and Lofgren's position as an aide gave him a front-row seat to the blurring of corporate and political agendas. After he retired, he said, "There is another government concealed behind the one that is visible at either end of Pennsylvania Avenue, a hybrid entity of public and private institutions ruling the country."

Similarly, Michael J. Glennon, a legal scholar, has reprised Bagehot's warning of "double government." People should not be deluded, he says, into thinking that Congress or the White House determines national security policy just because the Constitution says so. In reality, it is set by national security institutions that "operate largely removed from public view and from constitutional constraints." This is why US national security policy changes little, no matter who occupies the White House.

Glennon's concern makes sense when observing the uncanny continuity of foreign policies among the George W. Bush, Barack Obama, and Donald Trump administrations, three leaders whose views were vastly different. Obama's reproach of Bush's military adventurism abroad was a centerpiece of his campaign. Then he did an about-face once elected, shocking voters, by expanding the Afghanistan War with a massive troop "surge" and increasing the use of drones strikes and private military contractors abroad.

In due course, candidate Trump eviscerated Obama's meddling in Syria and Afghanistan, calling for the United States to withdraw from those regions, and quit NATO, too. Then President Trump flip-flopped on all three issues, stunning his supporters. Trump's only explanation was that things "are much different when you sit behind the desk in the Oval Office," which doesn't explain much. The deep state operates according to its own compass heading regardless of who is formally in power.

What's to be done? "Dissolve the unholy alliance between corrupt business and corrupt politics," urged Teddy Roosevelt a century ago. Eisenhower agreed midcentury, saying, "Only an alert and knowledgeable citizenry" can guard against the rise of misplaced power that "endanger[s] our liberties or democratic processes." Tellingly, this has not happened, or is it likely. If anything, the American deep state has only strengthened the alliance between corporations and politics. The landmark 2010 Supreme Court case *Citizens United v. Federal Election Commission* reversed decades of understanding by deeming corporate political contributions the same thing as individuals' free speech. However, no reasonable person would agree that a corporation, with its vast resources and single-minded agenda, is remotely the same thing as an individual person. Corporations are not people. But the double helix of corporations and politicos forms the DNA of America's power structure.

Deep states exist, and their naked power will become more apparent as states fade. Their unmasking will prove dangerous, as the protestors of the Arab Spring discovered. When a deep state is threatened, it does not go gentle into that good night. It attacks. It is one of the forces accelerating durable disorder, and through it other powerful countries will go the way of Iran, Turkey, Egypt, China, and Russia in the future.

New types of players will emerge as states decline, and some may grow into regional superpowers. This shift will fundamentally change who has leverage in international relations, upsetting world order later in the twenty-first century.

RULE 8: THERE WILL BE WARS WITHOUT STATES

Acapulco was once the most glamorous spot on earth. Hollywood flocked to this Mexican seaside town for its sun-kissed beaches, its deep sea fishing, and just to be seen. The ingénue Rita Hayworth celebrated her twenty-eighth birthday there with her husband Orson Welles aboard Errol Flynn's yacht. It's where JFK and Jackie honeymooned, and where Frank Sinatra hid from the mob. John Wayne and other stars bought the Los Flamingos hotel as their private getaway, inviting friends like Gary Cooper and Cary Grant to bask in the clifftop breeze. Zsa Zsa Gabor caused a sensation when she jumped naked into the pool. Elizabeth Taylor married her third husband, Mike Todd, in Acapulco, with Debbie Reynolds as her matron of honor. The town was "glorious," Reynolds recalls.

Today Acapulco is a battlefield. As with all war zones, there are plenty of bodies. In 2016, there were 113 killings per 100,000 residents, making it the third most dangerous city in the world. That's twice as much as the United States' murder capital, Saint Louis.

Acapulco is in the crossfire of a war without states. It is a transit point for US-bound Colombian cocaine and other narcotics, making it strategically important for warring drug cartels. Some might assume the Mexican government controls the city, but they would be wrong. For years, "Narcapulco" (as some call it) was controlled by the Beltran Leyva brothers. Anyone who got in their way was "disappeared"—so most didn't get in the way.

"What the Beltran Leyvas were doing was selling drugs," said Evaristo, a local who identified himself by only his first name, out of fear of reprisal. "But they left us alone."

The Beltran Leyvas worked for the powerful Sinaloa cartel, an international criminal syndicate. If a multinational corporation and a terrorist army married, their offspring would be a drug cartel. It can buy a kilo of cocaine in the highlands of Colombia or Peru for around $2,000 and sell it for upward of $100,000 in foreign markets—a 4,900 percent increase. And that's just cocaine. The Sinaloa cartel is both diversified and vertically integrated, producing and exporting marijuana, heroin, and methamphetamine as well. The CEO of this Narcotics Inc. is Joaquin "El Chapo" Guzman, a ruthless murderer with business savvy. He made *Forbes* magazine's list of top billionaires, and he is arguably the most powerful man in Mexico. The US intelligence community considers the Sinaloa cartel "the most powerful drug trafficking organization in the world."

Then the Zetas moved in. Imagine a band of SEALs going rogue and offering their firepower to a drug cartel. That's what happened in Mexico in the 1990s. Commandos from the Mexican army deserted and became muscle for the Gulf cartel, whose multibillion-dollar operations span four continents. Soon these enforcers broke away from the Gulf cartel and set up their own cartel, known as Los Zetas. Their hypercruel tactics of beheading, grotesque torture, and indiscriminate slaughter demonstrate their preference for brutality over bribery, extreme even by cartel standards. The US government calls the Zetas "the most technologically advanced, sophisticated, and dangerous cartel operating in Mexico."

A war between the reigning drug superpower and the rising one was inevitable, and Acapulco was the flashpoint. In 2006, a severed head was carried in by an ocean wave and deposited next to a Mexican sunbather and her two horrified children. It was one of six beheadings and scores of execution-style killings and grenade attacks that summer. Daylight battles took place in the streets between cartel hit squads, occasionally involving the police. One battle in the neighborhood of La Garita left flaming vehicles and multiple bodies.

"That's when all this began," Evaristo recalled, gesturing at shuttered shops and burned-out buildings.

Soon after, the Mexican government declared war on the cartels, and battle zones like Acapulco slid into World War III as dueling cartels and the military fought in the open. The military lost more often than it won, but in 2014 it captured El Chapo. However, he escaped prison through a mile-long tunnel. A year later, he was recaptured by Mexican marines in a shootout, and then extradited to a US prison. Since then, the Sinaloa cartel has splintered. Infighting between El Chapo's lieutenants has devastated places like Acapulco, while rival cartels carve up Sinaloan turf. The result has been more war.

Mexico's drug wars have dragged on for over a decade, its government powerless to contain the situation. Bloodshed spiked to record levels in 2017, and monthly homicide rates climbed to the highest in twenty years. New bodies appeared daily on the streets, with cartel firefights leaving twenty to thirty casualties regularly. Previously, violence had been concentrated in a handful of Mexican states; but in 2017, it spread nationwide, with twenty-seven of Mexico's thirty-one states recording an uptick in homicides compared with the year before.

"All the violence," said one Acapulco resident. "It's like being in Afghanistan or something."

When cartels go to war, the police are an afterthought. Most violence is cartel-on-cartel with occasional civilian deaths. This doesn't make it acceptable, but it does make it like any other war. Cruelty is a tactic in drug wars, similar to how it's used in terrorism. Severed heads and the tortured bodies of enemy combatants are left as warnings to members of enemy cartels. Sometimes those heads belong to police officers, cautioning authorities to back off, but narco-wars focus on defeating rival cartels, not the police.

Where is the Mexican government in all this? Sidelined as a minor actor. Its brazen declaration of war on cartels was a futile gesture that ended nothing. Cartels have since multiplied, and the fighting among them has intensified—all of which showcases the government's inability to govern. On paper, the government is the sovereign authority in Mexico, but in reality, the cartels rule. The only question is which one will dominate the country in the future.

For cartels, the state is a prize, not a force to be feared. Why kill your way to the top when you can just bribe? "Police and military are often complicit with drug traffickers," says one expert. "Huge quantities of drugs flow out of (and presumably cash flows into) areas where the military controls access." Corruption is not limited to individuals but is systemic, and officials who refuse bribes are killed. This is how cartels capture states (hence the term "narco-state"). Many Latin American countries are really just narco-states.

Sick of the government's impotence, some citizens have taken matters into their own hands. Cartels govern the countryside through intimidation, and they "tax" people via extortion. Farmers in the Mexican state of Michoacan, having had enough, took up arms and started killing cartel members. Within a few months, the vigilantes were mounting military-style raids with assault rifles, setting up cordons around towns and sweeping the area for narcos. Those they found, they killed.

"These people are filling a void left in their communities," observed one journalist. "The military's had thousands of soldiers here for more than seven years trying to take on the gangsters, and haven't really done anything. . . . It's really a damning indictment of a lack of effectiveness of the federal forces and the federal effort here."

Mexico is an example of a war without states, something we will see more of in the future. Drug cartels battle one another for control of the region while states are sidelined, or turned into zombie narco-states.

Unfortunately, modern strategists do not have a vocabulary to think about such conflicts, which is why they lose. After a decade of fighting, the Mexican government has failed to curb cartels, and violence has spiraled out of control. The United States has also failed, having invested forty years and $1 trillion fighting the so-called War on Drugs. Cartels have only grown stronger since Nancy Reagan's "Just Say No!" campaign. Why? It's not for lack of resources or political will. Rather, it's a failure of imagination about the nature of war.

Policy makers do not think of Mexico's drug wars as an actual war, despite the moniker "War on Drugs." For them, Mexico is like the musical *West Side Story*, plagued by dueling street gangs. Their solution is better policing, and that's the problem. They see the "drug war" as a law enforcement challenge rather than a real war. No wonder they are losing, decade after decade. Better police, new investigative techniques, and stronger laws will not win this new type of war. The solution is to reimagine war and change the way we think. Only then will solutions present themselves.

Here's what the drug wars teach us about modern conflict. First, why do we privilege some armed conflicts as war and regard others as crime? Mexico was the second-deadliest conflict in the world in 2016, but it hardly registered in the international headlines. As Syria, Iraq, and Afghanistan dominated the news agenda, Mexico's drug wars claimed 23,000 lives—second only to Syria, where 50,000 people died as a result of the civil war. The wars in Iraq and Afghanistan claimed 17,000 and 16,000 lives respectively.

Despite these facts, we still do not consider Mexico to be at war, which is absurd. Cartel hit men produce gruesome execution videos, just like ISIS, yet the world yawns. Narco-warfare is as bloody as terrorism and more of a threat to countries like the United States, but it is ignored because it's not considered war. The label "war" carries a strange legitimacy not accorded to other forms of armed conflict, and the international community musters significantly more political will and resources to ending wars than stopping a crime wave. No one ever got a Nobel Peace Prize for making an arrest. The irrational distinction between war and criminality is killing Mexicans daily.

Second, cartels are not street gangs but regional superpowers. The term "cartel" is a whopper of a misnomer. It implies business collusion and price-fixing, but nothing could be further from the truth. Mexican cartels fight it out. Moreover, they are much more than illegal businesses—they are drug empires. Their GDP is larger than that of many countries, grossing up to $39 billion annually, according to the US Department of Justice. If they were a collective country, its GDP would rank ninety-third in the world, ahead of Iceland and Bolivia. Their operations span continents, and unlike the legitimate businesses on the Fortune 500, they thrived during the 2008 recession.

Cartels are an example of the world's new superpowers. To defeat them, we must commit all elements of national power, not just law enforcement. It's what we do against terrorists and other lesser threats.

Third, when cartels wage war, they fight like empires. They battle each other for control of land, the resources on that land, and the people who can harvest those resources. It's pure exploitation, just as was done in the age of European colonial empires. Material wealth and martial conquest have long been a theme of war, from the Spanish conquistadors to the British East India Company. Merging the profit motive and war is nothing new, and cartels are one more example. In the case of Acapulco, the cartels fight for a strategic transit point. To defeat a cartel, we must use strategies of empire denial, such as containment, deterrence, coercive diplomacy, and military punishment. The Nazi empire was not defeated with a law enforcement mentality.

Lastly, why do we think of cartel muscle as thugs? Cartels operate through decentralized paramilitaries, which contain their own rank structure and internal discipline. Members of the lowest rank, known as *halcones*, or falcons, form the eyes and ears of the streets who spy on enemy cartel members and government security forces. *Sicarios* are the foot soldiers who conduct raids, ambushes, assassinations, kidnappings, thefts, extortions, and the defense of their *plaza* (turf). They are commanded by a *lugarteniente* (lieutenant), who is responsible for governing a parcel of land and maintaining discipline within the ranks. Cartel justice is harsh. At the top are the *capos*, or drug lords, who oversee the entire endeavor, like executives. They appoint territorial leaders, make alliances, and plan high-profile attacks—just like any king in history. These drug armies are more than a match for Mexico's military, or any military in Latin America. We need to employ strategies that combat paramilitaries, and this involves more than cops and robbers.

Lazy thinking is expensive—it costs lives, treasure, and international prestige. Yet it has dominated US drug war strategy for the past forty years, and it is why we continue to lose. Our insistence on treating cartels as substate criminal actors misses the larger point: we are facing a new type of warfare, one in which states are sidelined. Governments throughout Latin America have succumbed to cartels in these wars—and they are wars.

Criminal networks have replaced states in many parts of the world, and they will wage war. Acapulco is but one example. Believers in conventional war are blind to this, because these conflicts do not look like regular wars, and this blindness leaves us dangerously exposed. If we are to win, we must expand our strategic thinking to encompass wars without states.

REDEFINING WAR

Experts no longer know what war is. Buzzwords have replaced ideas, as authorities bicker over hybrid warfare, nonlinear war, active measures, and conflict in the "gray zone." There is no consensus about what these terms mean, other than that they refer to aspects of unconventional war. However, even this is dubious. As mentioned earlier, there is no such thing as conventional versus unconventional war—there is just war. "Conventional war" is a distinct type of warfare, just as "guerilla warfare" and "psychological warfare" are unique.

The only thing experts agree on is this: to be considered "war," an armed conflict must be fought for purely political aims, which is why narco-wars don't count. Fighting for material gain is somehow grubby and below war, making cartels criminals and a lowly police problem. "The use of force to curb criminal behavior such as piracy is not war," writes one expert, "because pirates seek material gains rather than political aims." This is the considered opinion of generals, scholars, and dictionaries. And it is wrong.

Material gains and political aims have long been entwined in war. Cartels are not the first. The Romans conquered their known world and grew filthy rich for a thousand years. The Mauryan Empire did the same in ancient India by marrying military strategy with economic theory, encapsulated in the book *Arthashastra*, which is still studied today, just not in the West. European powers colonized the planet from the 1400s to the 1950s—nearly six centuries—in their quest for gold, god, and glory. They considered this conquest war, as did the native peoples under their heel. The Opium Wars in China were about forcing European economic interests on a foreign people; the conflict had more than a few parallels to the modern drug wars. The cruelty of Spanish conquistadors is legendary, and England was said to have had an empire on which the sun never set and the blood never dried.

Cartels are no different. Narco-wars are not just about money; they are about holding territory and harvesting its resources, just like the colonial wars of the past. This requires governance, taxation (i.e., extortion), and strategy. Cartel methods are cruel, but so are those of states. The British colonized the Indian subcontinent with brutality, massacring civilians along the way. Their actions were so draconian that even their Indian soldiers rebelled in the Sepoy Mutiny of 1857, sparking a mass uprising that was gunned down cartel style. The United States settled its western frontier at gunpoint, using its army to drive off or kill Native Americans so that white settlers could take their land. In 1890, the Seventh Cavalry slaughtered between 130 and 250 Sioux men, women, and children at Wounded Knee, South Dakota. The American Indian Wars saw dozens of these massacres from 1830 to 1911. By comparison, cartels are more restrained.

People wage wars for many reasons, including to get rich. Is anyone really surprised? War experts are. Their selective memory ignores inconvenient history, though economists see it plainly enough. Obtaining wealth has always been an objective of war, on par with political aims. As the economist Milton Friedman asks, "Is it really true that political self-interest is nobler somehow than economic self-interest?" Believing wars are fought for purely political reasons is wrong.

The other big question that war experts get wrong is that of who gets to wage war. This is important, because the label "war" somehow connotes legitimacy; otherwise the conflict under consideration constitutes mass murder. Most experts assume that states—and only states—have the privilege of legitimately waging war, a Westphalian idea. Conflicts in which a state battles a nonstate actor are called unconventional wars, or "small wars," a derisive term dating back to the nineteenth century. Conflicts without states are not even considered war. Yet there are wars without states, as Acapulco shows.

War has moved beyond most war experts' understanding. Conventional war has become a relic, like a pay phone, and studies show that deaths in modern wars are overwhelmingly civilian. Many are true victims, but many are also combatants who do not wear uniforms or fight conventionally. Who is waging war? Not nation-states, as we saw in World War II. Rather, it is the world's new class of powers, and many of their conflicts form examples of wars without states. Narco-wars are one example.

Another example is the Rwandan genocide, a conflict that claimed 800,000 lives in ninety days. That's seven times the number killed in eight years of the Iraq War. This African war was not conventional in any sense. Traditional battle formations were nonexistent, and the main weapons were rape and the machete. Lucky victims could pay for a bullet. The laws of war were irrelevant, as was the United Nations. The belligerents were not states but rather two ethnic groups, the Hutus and Tutsis. The countries involved were states in name only. This kind of war is so alien to traditional warriors that they must affix a "state" label—Rwanda—to the conflict just to think about it, even though it wasn't about national interests and it spanned four countries: Rwanda, Burundi,

Uganda, and the Democratic Republic of the Congo. Amazingly, experts refuse to call this a "war," so powerful is their conventional war bias. Perhaps they think it was an anomalous 800,000-person homicide.

Traditionalists cannot contemplate wars without states, even though such wars surround us. Some are happy to ignore them, saying they should be considered just mass murder—a preposterous answer. Others struggle with ethnic conflict, not wishing to call it war but realizing it is more than murder. A host of prosaic terms exist to describe this netherworld of war, from "insurgency" to "war among the people." Most of the wars in Africa fall into this hazy category, and most of the world's wars are in Africa. Few war experts study Africa—a strange oversight.

Africa shows us the future of war. There are no conventional interstate wars there. Sometimes the state is a belligerent actor, such as with Sudan's war against the people of Darfur, which has exterminated 500,000 and burned whole villages. More often the state is a prize to be taken, as in the narco-wars. Several African warlords have become president this way, and they frequently remain in office for life.

Then there are wars in which states exist in name only. Conflicts in Somalia and the Central African Republic fall into this category, since those states do not really exist. We call them states only so we can find them on the map. The Congo region is home to another war without states. While the Second Congo War "officially" ended in 2003, the conflict rages on there to this day. The Congolese military is a faint actor there, and the United Nations rarely leaves its outposts, so who is waging this war? New kinds of powers. The Congo wars have also been the bloodiest on earth:

5.4 million dead, many times the death toll of the US wars in Iraq and Afghanistan. African wars make Middle East conflicts look like Boy Scout jamborees, but experts do not consider them wars, because there are no states involved.

War in the Middle East itself might be easier to comprehend if you remove states from the analysis. Who can forget images of ISIS blitzing across Iraq in 2014? Convoys of pickup trucks with big machine guns mounted in the back stormed across the desert. Terrorists dressed in black went door-to-door, interrogating by Koran. Those who could not recite key passages were shot in the back of the head. Some were shot regardless. The Iraqi army fled at the sight of ISIS's black flags, throwing down their weapons and ripping off their uniforms. What followed was ghastly: crucifixions, beheadings, defenestrations, sales of women and girls as sex slaves, and massacres of Shia.

To the West, this was an inexplicable horror. But to those involved, it was a war. In fact, it's an ancient war. Sunni and Shia Muslims have been at war over who would succeed Mohammad as the leader of the faithful since his death. This conflict has waxed and waned for 1,400 years, with ISIS as the latest player in this struggle. States are latecomers to this longstanding war, and many are just tools within the larger regional conflict, as are states in the narco-wars. That's why winning a single country, like Iraq or Syria, solves little in a war without states.

States are secondary in most Middle East wars, and the true belligerents are Sunni and Shia population centers that transcend national boundaries. The Shia are led by the ayatollahs in Iran, and their territory encompasses Shia populations in parts of Lebanon, Syria, Iraq, Yemen, and Bahrain—the "Shia Crescent."

Pushing back is a Sunni confederation led by the Saudi royal family; the alliance includes elements of the Gulf states, Jordan, North Africa, Pakistan, and Asia. Whole groups of people are involved in this war, ignoring their government's policies. Some countries, like Lebanon and Iraq, have sizable Sunni and Shia populations who often fight each other, sidelining their governments. Israel stays out of the way. On the eve of the Iran-Iraq War, another Sunni versus Shia slugfest, the Israeli Prime Minister Menachem Begin said, "I wish both belligerents good luck and much success!"

Because the conventional war mind sees only states, it will always lose wars without states because it cannot diagnose the problem. The front line in this Middle East war stretches from Israel across the Shia Crescent to Yemen. It is a single war, with many fronts. However, the conventional war mind views each country's conflict as a discrete war. Instead of making one strategy to combat a single war, it makes multiple strategies—one per country—and they work at cross-purposes. For example, the United States fights Iranian-backed Houthis in Yemen while fighting alongside Iranian forces against ISIS in Iraq. No one can win a war this way. This is not the US military's fault. It's a failure of strategic thinking at the top, and of the experts who advise the top.

What is war? War is armed politics, nothing more. Politics is not the sole province of states, and war can be waged by state and nonstate actors alike. It is waged for many reasons, in addition to purely political ones. People fight for economic gain, religious beliefs, identity, culture, glory, revenge . . . anything. And we must prepare for anything. There is no such thing as conventional versus unconventional war—there is only war. Think of warfare like smoke: always shifting, twisting, moving. Strategists who cling to rigid views of war will be blindsided by its mutable character, resulting in strategic surprise and defeat. As soldiers say, the enemy always gets a vote. To recognize the wars of the future, we must move beyond crude labels like "war," "criminality," and "mass murder." Such distinctions are fraught, because the ethical lines between them are blurry.

PRIVATE WARS

The rise of mercenaries coupled with new kinds of nonstate powers will produce private wars, an ancient form of warfare that modern militaries have forgotten how to fight. It is literally the marketization of war, in which military force is bought and sold like any other commodity. This will change warfare as we know it.

Privatizing war distorts warfare in shocking ways. If conflict is commoditized, then the logic of the marketplace and the strategies of the souk apply to war. A souk is an Arab open market, and a good analogy for how private wars work. In a souk, everything is up for sale and must be bartered. Anything goes. Fraud, deception, deceit, and hard bargaining are the watchwords. But so are value, rare finds, and exotic merchandise. Treasures are to be had, and for cheap—if one knows what one is doing. If not, expect to be scammed; this unregulated space is not for amateurs. There are no refunds, returns, or exchanges. Only street-savvy buyers should engage, and the best advice is also the oldest: caveat emptor. In the context of war, the implications are grave, as Machiavelli warns us.

What does private warfare look like in practice? It is the way of the mercenary, and soldiers for hire do not like to work themselves out of a job. The fourteenth-century Italian writer Franco Sacchetti tells a story that captures the perversion of private warfare:

Two Franciscan monks encounter a mercenary captain near his fortress. "May God grant you peace," the monks say, their standard greeting.

"And may God take away your alms," replies the mercenary.

Shocked by such insolence, the monks demand an explanation.

"Don't you know that I live by war," says the mercenary, "and peace would destroy me? And as I live by war, so you live by alms."

"And so," Sacchetti adds, "he managed his affairs so well that there was little peace in Italy in his times."

Privatizing war changes warfare in profound ways, and conventional strategists who fail to grasp this will get their troops killed. First, private war has its own logic: Clausewitz meets Adam Smith, the father of economics. For-profit warriors are not bound by political considerations or patriotism; in fact, this is one of their chief selling points. They are market actors, and their main restraint is not the laws of war but the laws of economics. The implications of this are far reaching. This introduces new strategic possibilities known to CEOs but alien to generals, putting us at risk.

Second, the fact of private warfare lowers the barriers to entry for war. Hiring mercenaries allows clients to fight without having their own blood on the gambling table, and this creates what economists call "moral hazard." Think of moral hazard like renting a car. Some people abuse the heck out of rented automobiles. Want to drive on train tracks at one hundred miles per hour? No problem. You would never do it to your own car, because it would cause long-term damage, but why worry if it's someone else's vehicle? You will never have to deal with the consequences, and this lack of personal responsibility encourages bad behavior in some drivers. It's the same with private war. Mercenaries are rented forces, and clients may be more carefree about going to war if their own people don't have to bleed. Mercenary captains might not care, either, if they do not have to fight themselves and instead order others into combat. Private warriors are disposable humans, similar to rented cars, and this fact emboldens recklessness that could start and elongate wars.

Third, private warfare breeds war. It's simple supply and demand, as mercenaries and their masters feed off one another. Here's how it works: Mercenaries and clients seek each other out, negotiate prices, and wage war for private gain. This prompts other buyers to do the same in self-defense, attracting additional mercenaries like ants to syrup. As soldiers of fortune flood the market, the price for their services drops and new buyers hire them for their own private wars. This cycle continues until the region is swamped in conflict, as it was in Machiavelli's day.

Private warfare's inclination toward escalation is a result of its economic nature. On the supply side, mercenaries don't want to work themselves out of a job. Instead, they are incentivized to start and elongate conflicts for profit. Out-of-work mercenaries become marauders, preying on the countryside for sustenance and artificially generating demand for their services. Sometimes they engage in racketeering and extortion of the defenseless. There is abundant historical evidence for this. "We find that our [mercenary] forces have cost the country a great deal and done much wanton damage," declared Frederick William, ruler of Brandenburg-Prussia, during the Thirty Years' War. "The enemy could not have done worse."

On the demand side, the availability of mercenaries means that buyers who have not previously contemplated military action can now do so. We've already seen multinational corporations, governments, and millionaires hire mercenaries in 2015; that was not the case two decades ago. The availability of private force lowers the barriers of entry into armed conflict for those who can afford it, tempting even more war.

Fourth, the market for force creates what political scientists call a "security dilemma." Think about an arms race between hostile countries that do not communicate well. During the Cold War, the United States and the USSR stockpiled vast amounts of nuclear weapons, mainly for defensive purposes. If America had one thousand warheads, the USSR wanted two thousand. The United States reciprocated with another three thousand, and then the USSR built an additional five thousand. Neither country felt secure unless it had more nukes. Ultimately this escalation laddered up to mega-arsenals, and one miscommunication could have blown up the world. In fact, it nearly happened at least six times during the Cold War.

Private warfare also creates a security dilemma. In such a dangerous environment, buyers retain mercenaries for purely defensive purposes, but this can backfire. Other buyers watch this amassing of force and suspect the worst—namely, a surprise attack—and procure twice as many mercenaries for their own protection. This prompts the first buyer, who also assumes the worst, to buy even more mercenaries, and soon an arms race ensues. The danger is that all sides will escalate and then unleash their forces. This lateral escalation creates a security dilemma, because people who do not wish to fight end up doing so anyway. Private wars invite more belligerents than public ones, and therefore increase chances of accidental war happening.

Fifth, double-crossing is the bane of private warfare. When mercenaries and their masters have a dispute, there are no courts of law to sue for breach of contract. Instead, things are settled by blood and treachery. Greedy mercenaries may wish to renegotiate their contract traitorously with violence, steal their client's property, or accept bribes from their client's enemies not to fight. Buyers who do not pay their bills may become victims of their own mercenaries, unless they hire a bigger mercenary outfit to chase them off. Because there are no laws in private warfare, market failure in this context means savagery.

Private warfare is the antithesis of conventional warfare, which is why modern militaries are unprepared for it. In fact, they cannot even comprehend it. To them, it's like wars without states: an impossible oxymoron. But this assumption is dangerously naive. Private warfare has been with us for millennia, even if it's now forgotten by modern strategists. In a free market for force, business strategies meld with military ones. In other words, private wars are driven less by politics than by political economy. Owing to this nuance, the conventional warrior will have problems identifying private wars, much less devising strategies to defeat them.

WINNING PRIVATE WARS

Not all wars without states will be marketized, but many will. Some stateless wars will be fought by ideological terrorists and insurgents, and they may turn to mercenaries for help as they become available. Below are some unique strategies to win such wars. Cunning is the watchword of private war, and readers should also consider the Thirty-Six Stratagems.

STRATEGIES FOR BUYERS (DEMAND SIDE)

- Bribe your enemy's mercenaries to defect.
- Retain all mercenaries in the area to deny your enemy a defense.
- Renege on paying mercenaries once they complete a military campaign.
- Give a larger mercenary unit a short-term contract to chase off or kill your unpaid mercenaries.
- Manipulate the winds of war by buying all the mercenaries available, driving prices up, then dumping them on the market, driving prices down.
- Engage in market defamation of specific mercenary units as a tool of accountability or blackmail.
- Rent new capabilities on the fly, such as a special forces team or attack drones, giving you maximum operational flexibility and unpredictability.
- If you have the money, outspend your rivals by waging an unlimited war of attrition. Mercenaries have a bigger recruiting pool than national armies, which are limited to their country's citizenry. The mercenary labor pool is global. This is especially useful when fighting a state committed to conventional war.
- Drive your adversaries into bankruptcy by stoking a mercenary arms race.

- Hire mercenaries as agents provocateur to draw others into a war of your choosing.
- Hire mercenaries for covert actions, maximizing your plausible deniability. This is useful for conducting wars of atrocity: torture, assassination, intimidation operations, acts of terrorism, civilian massacres, high-collateral-damage missions, ethnic cleansing, and genocide.
- Conduct false-flag operations: secretly hire mercenaries to instigate a war between your enemies, while keeping your name out of it.
- Hire mercenaries for mimicry operations to frame your enemies for massacres, terrorist acts, and other atrocities that provoke a backlash.
- Buy a large number of mercenaries, march them into your enemy's territory, and then release them, unpaid. Out-of-work mercenaries become bandits and will sow anarchy, accomplishing your mission on the cheap (unless your enemy hires them to attack you).
- Knowing the high danger of a mission, misrepresent it so that mercenary casualties will be extreme. Once they have achieved the mission, cut them loose and do not pay them. They will be too weak to challenge you.
- Hire multiple mercenary units to pursue the same objective without telling them. They will use different strategic approaches and sometimes work at cross-purposes. Reward the first unit that completes the mission and cut loose the rest, unpaid (hedging strategy).
- Hire multiple mercenary units to kill one another, thinning out their numbers and making them easier to control or swindle.

STRATEGIES FOR FORCE PROVIDERS (SUPPLY SIDE)

- Employ the shakedown strategy: blackmail or threaten the client for more money at a crucial moment.
- Start or elongate a war for profit.
- Negotiate and accept bribes from a client's enemies not to fight. Raise the price and offer to turn on your client, offering to stage a palace coup d'état.
- Bribe your enemy's mercenaries to defect, saving you battle costs.
- Secretly cut a deal with your mercenary opponents. Negotiate an outcome that benefits all mercenaries at the expense of clients.
- Engage in market defamation of clients as a tool of accountability or blackmail.
- Between contracts, become bandits for profit and artificially generate demand for protection services.
- Buy smaller mercenary units and incorporate them into your growing private army, giving you market power.
- Manipulate key military information that influences clients' business decisions in favor of your interests.
- Sell out your client to his enemy.
- Practice extortion and racketeering: Threaten to lay waste to a community unless it pays you protection money. Establish payments on an ongoing basis and raise prices whenever possible.
- Play multiple clients off one another to foster mistrust that leads to more war.
- Engage in Praetorianism: hold your client hostage and bleed

him dry of wealth for as long as possible. Look for a new host when finished.

- Establish a warlord kingdom to extract wealth from an area. This is especially useful in highly volatile regions rich in natural resources.
- Capture a high-value asset like an oil field or a small city and sell it back to its owner. When complete, ask for a contract to protect it from others like yourself.
- Steal your client's assets.
- Kill off your competition to become a monopolist and raise prices.

If these market strategies have left you queasy, you might be clinging to conventional war. Wall Street will recognize them as everyday business, and it may be more prepared to lead tomorrow's wars than today's generals are. The union of business ethics and the market for force is terrifying, but only a chump would deal himself out. War has moved on, and this is the future.

Private force will become a good investment as new consumers seek security in a deeply insecure world. New mercenaries will pop up to meet their demand, and so the market will grow. Expect future conflict markets to flourish in the usual global hot spots. However, introducing an industry vested in conflict into the most conflict-prone places on earth is vexing, since it exacerbates war and misery. Few would welcome an unbridled market for force, yet it is already here.

But this is not our worst problem. Another type of warfare is emerging, designed to crush conventional warriors and render their militaries useless. It's called "shadow war," and you are not supposed to see it.

RULE 9: SHADOW WARS WILL DOMINATE

A cold wind blows through a sprawling industrial district at the heart of Ukraine's war zone. A soldier scans the landscape through the scope of his Dragunov sniper rifle. All he sees is the same concrete wasteland: buildings pocked with bullet holes, streets full of bomb craters, and trees shredded by machine guns. There are no targets.

"Fuck it," he mutters in Russian.

Earlier that morning, automatic gunfire ripped through the winter air, but now the only sound is the creak of twisted, rusting sheets of metal. A new truce had been declared. Few believed it would hold.

Back at the base, a derelict warehouse, fighters huddle. All carry AK-47s, but none wear military insignia or a recognizable uniform. One man sits cross-legged on a cot and finds some comfort in the men's adopted dog, abandoned by its long-absent owners.

They use nicknames to disguise their identities. A man called Barmaley, after a fictional pirate and cannibal, carries a 1960s-vintage AK-47 with a homemade silencer. "When there is fighting to be done, those monitoring the cease-fire won't hear it," he explains.

Later that night, automatic gunfire erupts down the road, but none of the men stir. It is the battle rhythm of the front line: shooting at nightfall, and then off and on until dawn.

Miles away, Hammer is a gunner in the Donbass Battalion who fights for Ukraine. His military unit was formed by a private citizen, not the government. Some would call it a militia, while others would question who it fights for and why.

Hammer's squad is on patrol near the town of Marinka. Apartment buildings lay derelict, not a single windowpane intact. One has a five-story gash down its middle from a bomb, cleaving it in two. Hammer points to a disused stable two hundred meters away. It's where his enemy is holed up, a militia similar to his own but one that fights for the separatist Donetsk People's Republic. It takes orders from Moscow and gets shiny new weapons in return. The war in eastern Ukraine is actually a covert Russian invasion.

The men of the Donbass Battalion sneak around because they are badly outnumbered. A year earlier, their unit was trapped for a month at the Battle of Debaltseve. They were encircled by the enemy and took heavy losses.

"Twenty-five men dead from a company of eighty," he says back at the patrol base. Pulling out his phone, he shows film footage of the eventual rescue and evacuation. He and a friend, "Sniper," who has a ginger beard, are standing next to tanks in the freezing February cold.

"Russian soldiers took the town of Debaltseve," he says. "Not militia, but the Russian army. They didn't wear Russian rank or patches, but everyone knew who they were."

"Little green men" is what the world calls those men who won, named after the small plastic soldiers that children play with. Russia secretly deployed them to occupy parts of Ukraine and to take Crimea in a naked land grab.

"Spetsnaz, too," Hammer adds, referring to the dreaded Russian special forces. "Disguised as civilians." This is not a "laws of war" war.

Barmaley and Hammer were interviewed by the Western press, but their accounts hung low in a fog of conflicting reports. The West couldn't confirm that Russian forces were in Ukraine, so the world moved on. Vladimir Putin, the Russian president, denied the whole thing, of course, in a move harkening back to his KGB days.

"I will say this clearly," Putin said at a press conference. "There are no Russian troops in Ukraine . . . Russia is not going to try to annex Crimea."

Until it did. Once the annexation had been rubber-stamped by the Parliament in Moscow, Putin admitted that Russian troops had been deployed to Crimea after all. But the lie had served its purpose. The deception held the international community at bay while Russia mopped up the resistance in Crimea, incorporating it into what the Kremlin calls "New Russia." When pressed on these "little green men," Putin later insisted they were merely spontaneous "self-defense groups" who may have acquired their Russian-looking uniforms from local shops. When this became impossible to defend, he finally admitted they were Russian soldiers.

Years later, Crimea still remains a part of Russia, and the war in Ukraine grinds on, largely invisible to the world. That's by design. Moscow-controlled media organizations spin the facts at such a high RPM that even Russia experts are confused. The West will not risk a war with Russia if it cannot establish the basic facts of the conflict. It's a brilliant strategy by the Russians, in a diabolical sort of way. Meanwhile, Ukrainians are being plowed under. Over ten thousand have been killed, twice the number of American casualties in the Iraq War.

The Ukrainian conflict demonstrates how warfare has changed. In 1956, a student protest in Budapest led to a nationwide revolt against the subsequent Soviet occupation of Hungary. The same thing happened in Czechoslovakia in 1968. In both cases, the Soviets squashed dissent under their tank treads.

That was the twentieth century, when military might was power. Now it's a liability. Russia had enough tank divisions to blitzkrieg Ukraine but instead chose to conduct covert operators. Why? Because clandestine forces allow Russia to escalate the conflict in secret. Little green men, proxy militias, mercenaries, and Spetsnaz flooded the countryside, forming a ghost occupation force. By the time the international community figured it out, Russia's conquest was a fait accompli.

Plausible deniability is more decisive than firepower in the information age. How can the United States or the United Nations rally the world to fight a war that may not exist? They can't. It's an effective strategic offense by Russia, and it's an example of what's to come.

THE SHADOW IS MIGHTIER THAN THE SWORD

War is going underground and will be fought in the complicated shadows. Militaries can no longer kill their way out of problems when everyone is armed with high-resolution video-taking mobile phones that can upload content to the 24/7 news cycle from anywhere. Nations, it seems, can suffer casualties but not bad press. The earlier rules described in this book showed what doesn't and won't work in modern and future wars. This rule shows what does.

Shadow wars are armed conflicts in which plausible deniability, not firepower, forms the center of gravity. This dynamic makes war epistemological: telling what is real from fake will decide the winners and losers. Don't expect large tank battles. Warriors will be masked and offer good plausible deniability, as offered by special operations forces, mercenaries, terrorists, proxy militias, little green men, and foreign legions. Black ops will be the only ops that matter, and those who fight according to the laws of war using conventional war strategies will have those strategies used against them. Terrorists and others routinely use this method, and it's hard to understand why the West has not updated the laws of war or opted to fight a different way before sending its young men and women into gunfire.

In a shadow war, cloaking is a form of power, and information is weaponized. If you twist your enemy's perception of reality, you can manipulate him into strategic blunders that can be exploited for victory. It's also a great defense. Near the beginning of the Ukraine war, Russian proxy forces blew Malaysia Airlines Flight 17 out of the sky, killing all 298 people on board. It was the deadliest episode involving an airliner being shot down ever, and the ninth-deadliest disaster in aviation history. The wreckage scattered across the Donetsk region, which was initially cordoned off by Russian forces to prevent outside investigators from learning the truth.

A battle of narratives followed. The West blamed Russian separatists inside Ukraine for firing a Buk surface-to-air missile, whereas Russia claimed that Ukrainian forces had fired the missile. The United Nations demanded a full inquiry, but armed Russian separatists blocked access to the crash site, so the UN gave up. The world erupted in outrage until a new scandal seized the news cycle, and everyone moved on. Without clear evidence, it's hard to know truth from fiction. War is becoming a "he said, she said" affair with no meaningful consequences for liars.

Russia has become a disinformation superpower, employing a "kill 'em with confusion" strategy. And it's working. The evidence is everywhere: making a war in Ukraine invisible, hacking the 2016 US presidential election, stoking the Brexit vote, supporting fringe political groups, fueling right-wing nationalism in NATO countries, and spinning its dubious role in the Middle East. Since 2010, the Russian military has prioritized what it calls "information confrontation" to guarantee information superiority in peacetime and wartime. Russia is now an empire of lies.

We all think of ourselves as savvy media consumers, yet Russia succeeds in manipulating public opinion. How does it do it? The answer is found in Milbank Tower, a slick skyscraper a few blocks from London's Westminster Palace and Thames House, the MI5 headquarters. There, journalists beaver away amid state-of-the-art equipment in glossy studios, using slick graphics to pump out the truth, Russia style. Similar sites exist across the world. They all work for the RT news network, and their programs look and feel like those produced at CNN, Fox, the BBC, and France 24. RT features programming tailor-made for US and UK viewers, as well as offering services in French, Spanish, and Arabic. Its reach is global, broadcasting to some one hundred countries via satellite television and the internet. But no one should be fooled into thinking RT is a news outlet.

RT is not a media company but an intelligence operation, and its purpose is not information—it's disinformation. It offers "alternative facts" to seed doubt and change minds. Taking a page from the embattled spook's handbook, its mantra could be: "Admit nothing. Deny everything. Make counteraccusations." The Kremlin funds RT's $400 million annual budget to warp the truth for Russia's strategic interests. Its spies even have a name for this kind of subversion—"active measures"—and it's an example of how shadow wars are fought by weaponizing information. One reason why RT is effective is that it blends legitimate experts and journalists with crackpots, offering a plausible version of events that is nested within a larger global disinformation campaign. Think of RT as strategic storytelling.

The "Troll Factory" is another component of Russia's active measures against the West, revealing the true power of cyberwarfare. It's not sabotage, like Stuxnet—it's disinformation. Located in Saint Petersburg and officially called the Internet Research Agency, it's where Russian operatives hack into websites, create phony news sites, and pump out fake news and bogus social media messages. Trolls are anonymous agents provocateur that stalk the internet, throwing seditious hand grenades into chatrooms and on news sites. Then there are bots, programs that mimic trolls by the thousands and drown out legitimate content. The West has few defenses against this subversive blitzkrieg.

The Troll Factory's mission is to manipulate Western public opinion to serve Russian interests. For example, it dropped 45,000 garbage tweets on the United Kingdom during the final forty-eight hours of the Brexit referendum, and some believe it altered the close vote. Russia wants to explode the European Union, and Brexit could be the spark that ignites the fuse. Russia also likes to disrupt democracy. The CIA, the FBI, and the National Security Agency all agree with "high confidence" that Moscow tried to swing the 2016 US presidential election to Donald Trump. The Department of Justice investigation led by special counsel Robert Mueller has found evidence for this, charging thirteen Russians and three Russian companies, including the Internet Research Agency, of having "a strategic goal to sow discord in the U.S. political system, including the 2016 U.S. presidential election." Congressional hearings, multiple investigations, media inquiries, public outrage, and White House scandal has frenzied the country ever since, as Putin chuckles.

Globalization makes the size, speed, and scale of strategic disinformation more powerful than in any time in the past. Democracies are especially vulnerable to active measures because of what political scientists call the "CNN effect." People see compelling images on TV or the internet, such as pictures of a humanitarian crisis, and then demand that their leaders "do something!" Politicians looking for votes then order a humanitarian intervention, even though it's not in the national interest to do so. This is not new, and it dates back to at least the sinking of the USS *Maine* and the Spanish-American War. RT and the Troll Factory exploit the CNN effect by cranking out "alternative truths" to sway elections and make the Ukraine conflict invisible. One can imagine Putin with a Jedi hand wave: "These are not the little green men you're looking for." By the time Russia seized Crimea by force, it had already won the shadow war. Crimea was not a battle—it was booty.

Shadow war is powerful because it weaponizes information in an information age. Conventional warriors shrink from the media spotlight, whereas shadow warriors embrace it. Having the 24/7 news cycle catch your troops committing war crimes spells defeat for the conventional warrior. Not for the shadow warrior. Not only does he deny they are his troops but claims they belong to his enemy, who must be sent to the Hague! For shadow warriors, the media is not a liability but an opportunity. The conflict in Ukraine is just one example of how shadow strategies defeat conventional ones in modern warfare.

THE DARK ARTS

Subversion will be everything in future wars. Who cares how many nukes you have if you don't know where to point them? Subversion renders blunt force secondary, as China's Three Warfares strategy demonstrates. Russians call their version *maskirovka*, or "masquerade," and it has been a part of its strategic culture since the fourteenth century. What began as crafty military deception is now the Russian way of war.

Maskirovka's strategic logic is compelling. It manufactures a fog of war and wins by turning the enemy into a sock puppet. Such dark arts are the true weapons of mass destruction, not nukes. For example, Russian active measures could corrupt intelligence databases, analysis, and conclusions. Why invade a country when you can trick the West (or someone else) into doing it for you? This is shadow war.

Deception is the oldest form of warfare, and it is the most formidable. Sun Tzu wrote 2,500 years ago, "All warfare is based on deception," and he explained how to employ such strategies in his book *The Art of War*. This text is one of the supreme classics on strategy, useful for war, business, politics, sports, family . . . you name it. Too often the West misreads *The Art of War* as the fortune cookie of strategy, due to its concise nuggets of wisdom and the many poor translations that have been made. You also need to know a little about ancient Chinese Taoism and the cosmos of the *I Ching* to understand Sun Tzu. Sometimes my war college students joke that Sun Tzu is the Yoda of warfare. Or a dark lord of the Sith.

Sun Tzu advises the "indirect approach" to war, a strategic idea taken up briefly by the West after the calamity of World War I but then ditched. It comes down to this: don't fight your enemies—outfox them. Done well, this approach manipulates the enemy in order to create vulnerabilities you can exploit. Unlike Clausewitz, Sun Tzu thinks force is the fool's way of war, and battlefield victory the mark of an inept general. The zenith of skill is to trick your enemy into losing before he even comes to battle. "The supreme art of war," he says, "is to subdue the enemy without fighting." Wit beats muscle.

The indirect approach requires a few things to succeed. Here are just two. The first is information supremacy. You can't outsmart your enemies without knowing everything about them, and Sun Tzu was obsessed with espionage. He even lays out a classification of spies, including ones for sacrifice. But success also rests on grasping your own capabilities and limitations; hence his maxim "Know your enemy and know yourself. In a hundred battles you will never be in peril."

The second is knowing that a cunning mind is superior to a martial one. Don't rush headlong into a firefight with your enemies. Instead, bait them into a firefight among themselves and mop up the survivors. Use deception to create chaos, then exploit it. When capable, feign incapacity; when active, inactivity. When near, make it appear that you are far away; when far away, that you are near. Fake disorder, then strike an unprepared enemy. Deception wrong-foots your enemy and keeps him guessing. The Thirty-Six Stratagems also utilize subterfuge.

Needless to say, this is not the West's way of war. Western militaries revere Clausewitz while Sun Tzu is rarely taught (and when he is taught, he's rarely taught well). It's hard to imagine what a dinner party with the two theorists would be like. Clausewitz thinks brute force and battlefield victory is everything; for Sun Tzu, it's nothing. Clausewitz curses chaos and "the fog of war" as barriers to victory; Sun Tzu creates chaos and weaponizes it for victory. Clausewitz believes cunning ruses are the weapon of the weak; for Sun Tzu they are the weapon of choice. Clausewitz thinks spies untrustworthy and intelligence reports unreliable; Sun Tzu finds them indispensable. Clausewitz is the father of "conventional" war; Sun Tzu is the father of "unconventional" war . . . or whatever we're calling it these days. Clausewitz is the legionnaire; Sun Tzu the ninja. Clausewitz is the lion; Sun Tzu the fox. No doubt there would have been broken china on the floor by dessert.

In shadow war, subversion is the strategy and plausible deniability the tactic. Rather than fight the forces of durable disorder, shadow wars harness them by creating chaos and using it. In other words, the essence of shadow war is to keep the enemy guessing. The Thirty-Six Stratagems offer some ideas on how to achieve this, and in all of them, cleverness wins over brutality. The shadow warrior is skilled in attack if the enemy does not know what to defend, and an expert in defense if the enemy does not know what to attack.

INSURGENCY VERSUS SHADOW WARS

Shadow wars may mimic or hijack insurgencies, and this causes problems for strategists. Russia's annexation of eastern Ukraine is an example, as it poses as a pro-Russia insurgency. It's important to distinguish between an authentic insurgency and a shadow war because they require different responses. Mistaking one for the other plays into the shadow warrior's ruse. The first rule of shadow war is: Don't be a sucker.

The best way to distinguish between an insurgency and a shadow war is to examine how the local people are treated. Insurgents, even ruthless ones, need the people. Sometimes it doesn't seem so, and many insurgencies have failed because of excessive violence against the local population. This happened to al-Qaeda during the Iraq War. The terrorists initially won over local tribesmen but treated them so cruelly they switched sides and fought for the Americans, in an episode known as the Sunni Awakening. A better model of insurgency is the American Revolution, in which the Founding Fathers knew they had to treat the population with respect because they would be future citizens, empowered with a vote.

In shadow wars, people are prey. Civilians are more than collateral damage; they are useful military targets. This is especially true when fighting an enemy who cares about the population, such as an insurgency or military that wishes to respect human rights. In a Sun Tzu move, the shadow warrior uses this noble desire against his enemy. However, it works only if you have plausible deniability, the shadow warrior's weapon of choice.

Slaughtering the innocent is used in such wars to bait, punish, or provoke, and it works. It even works with one's own people. When Putin came to power in 1999, he faced a fractured nation being devoured by organized crime and anarchy. Then he had his moment. A series of explosions hit four apartment blocks in three Russian cities, including Moscow, killing 293 and injuring more than 1,000 people. Soon it was discovered that Chechen terrorists were behind them, and waves of panic swept through Russia. It was that country's 9/11, leading to the Second Chechen War, a brutal affair that united the nation in common cause.

Except it wasn't the Chechens; it was Putin seeking to consolidate his political power, facilitating his ascendency from the prime minister's office to the presidency. Experts now know that he and the FSB (the agency that succeeded the KGB) were behind the bombings. John Dunlop and Amy Knight, specialists on the FSB, give ample evidence showing that the FSB was, in Knight's words, "responsible for carrying out the attacks." She concludes: "It is inconceivable that it would have been done without the sanction of Putin." Of course, those familiar with Putin's career are hardly shocked.

Killing civilians to manipulate the winds of war and to achieve indirect strategic effects is the shadow warrior's way. They will get away with it in the future because the laws of war have devolved into a punch line. Warriors are increasingly masked and kill indiscriminately. Rules of engagement are nonexistent, and civilians are targets. Conflicts in the Middle East, Africa, and Afghanistan are replete with human rights violations, and the international community does little to stop them. These are all signposts marking the rise of shadow wars, which will be the dominant form of warfare in the decades ahead.

BITTER FRUIT

The West wasn't always so pathetic in the practice of the dark arts. In fact, it was quite good at it, in a morally challenged way. In 1950, Guatemala elected a president who promised to alleviate crushing poverty, and he did. Briefly. President Jacobo Árbenz enacted landmark agrarian reform that benefited 500,000 people by redistributing land from the top 1 percent to everyone else. Perhaps he could have gotten away with it, too, had the United Fruit Company not been one of the evicted landowners. United Fruit (now Chiquita Brands International) was an American multinational corporation more powerful than any state in Latin America, and it had effectively colonized the Caribbean with "Banana republics."

United Fruit had friends in high places. In the past, it had persuaded the US Marines to invade countries of its choosing, such as Honduras, in order to boost its profit margins. In the first half of the twentieth century, the Marine Corps was reduced to a gang of Wall Street cronies in another example of the American deep state at work. General Smedley Butler, a two-time Medal of Honor winner who saw action throughout the Banana Wars, complained he had become "a high class muscle man for Big Business, for Wall Street and the bankers . . . a racketeer, a gangster for capitalism." Not this time, though. The marines would not invade Guatemala for United Fruit's bottom line.

Not dissuaded, United Fruit turned to Edward Bernays, the original "mad man" of Madison Avenue and the guy who coined the term "spin" to describe the public relations business. Bernays was a master of the dark arts and had come to United Fruit's rescue before. He reached out to two lawyers who once worked with United Fruit, John and Allen Dulles. Fortunately for him, the Dulles brothers ran the State Department and the CIA at the time. Bernays convinced them that a Communist insurgency had overrun Guatemala, and that something must be done. Something was done.

The Dulles boys persuaded President Eisenhower that Árbenz was a Communist puppet and needed to be removed in order to prevent a Soviet takeover of Central America. "Guatemalans are living under a Communist-type reign of terror," asserted Secretary of State Dulles. Worse, Árbenz's government would encourage the USSR to put nukes on Texas's border, giving it first-strike advantage. America would never survive such an attack. The Dulleses insisted that the United States find a way to roll back communism immediately, before it was too late. Eisenhower agreed.

However, they had a strategic problem. The United States could not invade Guatemala outright, because it was a presumed Soviet client state, and any hostile action against it could provoke World War III. The Dulleses had to find another way to depose Árbenz that gave the United States plausible deniability. The CIA came up with a Sun Tzu solution, code-named Operation PBSUCCESS. Its secret mission was "to remove covertly, and without bloodshed if possible, the menace of the present Communist-controlled government of Guatemala" and "to install and sustain, covertly, a pro-US government."

In the summer of 1953, the CIA began its shadow war against Guatemala. Operatives covertly recruited students to paper Guatemala City with anti-Communist stickers, and they enlisted American pilots to buzz government facilities and drop leaflets, intimidating locals. In the months ahead, the CIA manufactured consent through directed rumor, pamphleteering, additional poster campaigns, graffiti, and intimidation. The CIA also created a fake guerilla radio station called Voz de Liberación ("Voice of Liberation") just across the border in Honduras. It broadcast daily to the Guatemalan "resistance," complete with rousing music, giving the impression that an insurgency was swelling inside the country.

To bolster this deception, the CIA installed a rival of Árbenz's, Castillo Armas, a former military officer who earlier had tried and failed to overthrow him. Armas's forces were staged in bases in Honduras and El Salvador, their numbers hugely exaggerated by CIA propaganda. Meanwhile, Secretary of State Dulles instructed the US diplomatic corps to isolate Guatemala politically, giving Árbenz the impression that no other country would come to his aid when the Americans invaded.

Shortly before the supposed invasion, the US Navy set up a blockade and warplanes began to practice bombing runs. The Guatemalan people now believed an invasion was imminent and began an exodus. CIA propaganda convinced the Guatemalan military to turn on Árbenz and save their country. It worked. Árbenz fled for his life, and the country was safe once more for United Fruit and American interests. The CIA's postmission report concluded that Operation PBSUCCESS "had been accomplished and plausible denial retained." Putin would have been proud.

The CIA's shadow war against the Árbenz regime was a victory. Using Sun Tzu strategies, it achieved a bloodless coup and saved thousands of lives, compared with the expected cost of a military invasion. However, the greater triumph goes to United Fruit, which manipulated the CIA from the start. One of the perennial problems of shadow wars is knowing whom you're working for and why. Deception is everywhere, and the CIA had taken Butler's place as a Wall Street stooge.

What followed in Guatemala was ghastly, and though it has little to do with shadow wars, the tale bears telling. After Árbenz left, the CIA installed Armas as president in 1954. Not happy with him, the agency backed a series of military dictators, each worse than the last, until the whole country became embroiled in a civil war that lasted twenty-six years. During that time, over 200,000 civilians were killed, and atrocities were common, including massacres, rapes, and disappearances. The United States' complicity in some of this remains debated.

Congress finally had enough of the CIA's shenanigans. It wasn't just Guatemala, either, it was everything: the Iranian coup (a shadow war like that in Guatemala the same year); the failed Bay of Pigs invasion; its possible complicity in the JFK assassination; the Gulf of Tonkin Incident, which led to the Vietnam War; its infiltration of US domestic civil rights groups; its mucking around in Chile; and its role in Watergate. The CIA's shadowy manipulations had grown out of control, and the 1970s became the decade of intelligence reform. Congress's Church and Pike Committees excoriated the agency, as did the press. President Carter purged eight hundred covert-operations positions on October 31, 1977, in what became known as the Halloween Massacre. The CIA was never the same. But it was necessary. Its operatives had bungled themselves to death.

The lesson here is not that shadow wars don't work—they do—but that secrets and democracy are not compatible. This means democracies will be disadvantaged in an era of shadow warfare, a fact Putin already exploits. Democracy thrives in the light of information and transparency. Shadow wars favor the darkness of autocracy. Unfortunately, some democracies may be tempted to sacrifice their values in the name of victory, a phenomenon that's as old as democracy itself. The ancient Greek historian Thucydides watched Athens become increasingly despotic as it fought its rival Sparta, an authoritarian regime, during the Peloponnesian War. By the end of the war, Athens was no different from Sparta, and it lost anyway.

Democracy dies without transparency. The West can try to shed light on shadow wars and expose them, but this is not enough; it didn't save Crimea or the passengers on Malaysia Airlines flight 17. The West needs to learn how to fight in the shadows without losing its soul, or it will continue to get sucker punched by autocracies.

WESTERN SHADOW WAR

War is going underground, and we must go with it. Many conventional warriors will reject a Western version of shadow war, but this cannot be helped. Their strategies have failed us for decades and now play into the hands of shadow warriors, causing us to lose. Others think that using subversion is somehow morally wrong. A hundred years ago, these people probably would have objected to the machine gun, the submarine, and the bomber as unsporting. Warfare is ever changing, and we must adapt or die.

The bigger difficulty is that shadow wars harm the soul of a democracy. But kneeling before dictators is not an option. The West must develop its own version of shadow war, and it should not mirror that of Russia, China, or Iran. Instead, it can be tailored to undermine autocracies, which is easier than undercutting democracies. The West also has unique levers of power that can be utilized. Here are some weapons a democracy can use to fight back against autocratic shadow warfare.

The first set of tools are kinetic, meaning guns and men and women to shoot them. Specifically, these kinetic tools consist of nonattributable forces that are designed for maximum plausible deniability, conducting "zero-footprint" missions that are invisible to the world, and especially to the target. Such forces can also perform misattributable or false-flag operations that frame foreign actors. In the future, it's a frame-or-be-framed battlefield.

The shadow warrior class includes special operations forces, the foreign legion, proxy militias who fight for shared interests, masked soldiers (or "little green men"), and mercenaries of every stripe. Battles will become black-on-black affairs. Needless to say, these forces will not fight conventionally or be bound by the laws of armed conflict. Such strictures are from a bygone era, and modern war has left them behind.

The next set of tools are nonkinetic. Weaponized information will be the WMD of the future, and victory will be won in the influence space. The objective is to manipulate the enemy's decision-making calculus and sap its will to fight. To accomplish this, the West must develop its own active measures to gain information dominance. Myth-busting alone is insufficient. Setting the record straight is not enough to dispel the spin of Russia, China, and terrorists. Strategic influence is not the genteel art of debate. Instead, it is aggressive and devious, and it has to be. In poker, there is an adage: If you can't spot the chump at the table, then you're the chump. Too often, the West is the chump. It must overcome its aversion to knowledge manipulation and figure out how to fire nonlethal weapons.

The mantra of active measures should be "To inform is to influence." Examples include manufacturing dissent through trolls and bots, thereby leveraging the true power of cyberwar; clandestinely supporting dissenting voices in the enemy's camp; and establishing front organizations to push one's agenda or counter an enemy's narrative. These tools are useful for defense, too. For example, if an enemy is covertly sponsoring a nonprofit think tank to stir up opposition in your nation's capital, don't waste time officially denying it. This will only legitimize their cause and attract the attention of Big Media. Instead, create a "grassroots" group or rent a competing think tank to contradict its message. To the outside world, it will look like two public policy institutions having a catfight, and viewers will pass it by with a yawn.

Autocracies have developed active measures to disrupt democracies, so the West should return the favor. Ridicule works exceptionally well against autocrats, who must manufacture a cult of personality and fear to stay in power. Democracies should "un-manufacture" it. Seek to corrupt the enemy's trusted information sources and its intelligence agencies. The Russian general staff is said to have a hive mind. "Unhive" it by sowing uncertainty, causing its officers to make bad decisions that can be exploited. Over time, this uncertainty may even catalyze internal dissent, cracking the edifice of autocracy. Counter terrorist ideologies through covert denigration campaigns that make the world laugh at the terrorists' cause. When dealing with anti-Western regimes like Tehran, bomb them with Barbie dolls. Or deploy Homer Simpson.

Autocrats are vulnerable to paranoia because their rivals are everywhere, ambitious, and deadly. The West should exploit this. For example, one could conduct a whisper campaign that pits elites against one another, and watch the enemy purge its own ranks. Think of Stalin and Saddam Hussein, who decimated their militaries and power elites for fear of disloyalty. The West might secretly back opposition parties or factions to encourage regime change from within. Or it should support regime changes like Ukraine's "Orange Revolution," which deposed a Russian puppet. Moscow fears these "color revolutions," calling them a new form of warfare, so the West should use them to apply pressure.

For example, Russia can focus on the Middle East only because it is undistracted by pushy satellite states. It's time that the West started supporting those satellite states again, as we did in the Cold War. Get Moscow worried about its home front by facilitating underground color revolutions, and it will pull out of the Middle East. Bombers cannot achieve this.

Other tools hit the enemy's pocketbook. In the old rules of war, this meant sanctions and other blunt instruments that were never as effective as promised. Sanctions starve only the masses, not the elite who make the decisions. The plump leader of North Korea is a case in point. Shadow war takes a different approach. What about sponsoring criminal activity? Organized crime can become the enemy within the enemy's state, and you can shape the behavior of the criminal organization at the same time. For example, demand it stop trafficking human beings and end all sex trade activities as a condition of sponsorship.

One can facilitate kleptocracy and corruption within an enemy's camp to erode effective governance and challenge an autocrat's hold on power. When facing oil-dependent petro-economies like Russia or Iran, find ways to crash the price of oil, ruining their economies and making these countries more compliant with your demands. Make it easier for multinational corporations to do business in areas that threaten your enemy's interests. Often these corporations can go places government officials cannot, and open doorways into denied areas. More importantly, their activities can create webs of "sticky power" that make it hard for an autocrat to move against you because of the specter of mutually assured financial destruction. Laws like the Foreign Corrupt Practices Act should be reconsidered, as they already put countries like the United States and the United Kingdom at a strategic disadvantage in the global marketplace, where facilitation fees (a.k.a. bribes) are business as usual.

Lastly, apply pressure by sanctioning autocrats' friends and family members. Let them change the dictator's mind. For example, ensure their credit cards never work overseas and freeze their illicit bank accounts, curtailing million-dollar shopping sprees in Paris. Prevent their children from receiving student visas to study at Harvard and Oxford. Some may think this is unfair, but it's not. These kids dine in opulence ripped from the backs of the downtrodden, and they will inherit their fathers' kingdoms and vile ways. For example, one of the sons of Muammar Gaddafi received his PhD at the London School of Economics with a dissertation extolling the virtues of democracy. Five years later, he joined in his father's brutal response to the democratic uprising during the Arab Spring. Many of the West's top schools shamefully admit the children of the global 1 percent in exchange for a bribe. Perhaps this corruption cannot be stopped, but it can be weaponized.

Diplomacy also matters, but not in the way people think. Diplomatic statecraft evolved with the Westphalian system and is dying with it. Ministries of foreign affairs and the US State Department were designed to talk to other states, but now nonstate actors eclipse many countries in power. The institutions of a state must communicate with all actors of relevance, not just the ones with flags. This includes multinational corporations, terrorist groups, and criminal organizations who exert influence. Conventional diplomats will reject this, of course, but is engaging with such actors any different from dealing with distasteful regimes? Some of these nonstate actors will wage war in the future, and alliances with them will prove obligatory when confronting a common enemy. This is coalition shadow warfare.

The strategic possibilities are endless for the cunning mind. The key is this: all these tools are mutually reinforcing and inexorably linked—nothing occurs in a silo. For example, disinformation campaigns must support covert operations on the ground, and vice versa, otherwise the strategy will fail. Actions die in isolation.

Shadow war is an effective strategy for the future, and the West should adopt some version of it. Sun Tzu told us 2,500 years ago that all warfare is deception—and it still is. In fact, it's more so now. Modern war leverages information because we live in an information age. This is why it's effective. Power no longer comes out of a barrel of a gun, but rather from the complicated shadows.

RULE 10: VICTORY IS FUNGIBLE

In 1917, the February Revolution broke out in Petrograd (the name given to Saint Petersburg at the beginning of the First World War). Thousands took to the freezing streets to protest food shortages, corruption, and the disastrous war against the Central Powers. Police officers tried to beat them back but were swarmed by the starving masses. Unrest soon spread to other parts of Russia, and soldiers loyal to the czar marched with the people. Fearing for his life, Czar Nicholas II abdicated and was arrested.

In Zurich, a young Polish revolutionary bolted up the stairs to a squalid one-room apartment. Vladimir Lenin and his wife had just finished lunch.

"Haven't you heard the news?" the man shouted. "There's a revolution in Russia!"

Celebration ensued, as Lenin and other dissidents cheered their good fortune. This was the opportunity they had hoped for. Soon, they thought, Russia would be a worker's paradise. However, they had a big problem: because they were Russian citizens, all routes between Zurich and Petrograd would be closed to them by the Central Powers. In despair, Lenin feared history was passing him by.

In Berlin, Richard von Kühlmann sensed an opportunity. He was Germany's secretary of state for foreign affairs and its chief diplomat. The Great War had brought the world powers to their breaking points. The quick victory promised by the Schlieffen Plan had failed, and now Germany was caught between two pincers: England and France to the west, and Russia to the east. Both fronts were stalemated, devouring millions of men and hemorrhaging resources. Kühlmann knew Germany would not last the year unless something changed. Surrender was not an option. Germany had captured nearly half of France; why should it be the one to yield?

The only path to victory lay in quickly winning one of the fronts, allowing Germany to concentrate all its might against the other. That hadn't seemed likely, until now. The German high command wanted to throw millions more men at the problem but had none to spare. As the inheritors of Clausewitz's legacy, they could think only in terms of force and attrition. But Kühlmann was a businessman turned diplomat and saw things another way.

German agents approached Lenin in Zurich and offered him the deal of the century. Germany would secretly transport him and his followers in a "sealed train" across Europe and into Petrograd. A sealed train is like a huge diplomat pouch: it travels across borders under an immigration seal, and customs officials are forbidden from looking inside.

A week later, Lenin and his coterie arrived in Petrograd like a plague virus, infecting Russia with their ideology and knocking the country out of the war. A year later, Russia's exit was formalized with the Treaty of Brest-Litovsk. In a single train ride, Kühlmann had taken out the eastern front, not with bullets, but with cunning. As he explained, stripping the Allies of an ally was "the purpose of the subversive activity we caused to be carried out in Russia behind the front." Kühlmann won by weaponizing Lenin's virulent ideology.

Lenin's sealed train illustrates that there are many ways to win in war—in other words, that victory is fungible. For the unimaginative members of the German high command, victory could be achieved only on the battlefield, following a "he who has the most bullets, wins" logic. But they were conventional warriors. Kühlmann shows us an alternative approach. It was cheap to execute and easy to accomplish, and it saved millions of German lives, too. Such a stratagem would never occur to conventional warriors, because they understand only the language of violence. Violence has its place in armed politics, but there are many languages of power. Military might is but one, and it is not always the most effective.

The war did not end well for Kühlmann. Germany was too late in closing the eastern front, and the United States had just joined the war, tipping the scales of victory toward the Allies. Wars of attrition are won by body count. In July of 1918, Kühlmann went to the front and pleaded with the German kaiser Wilhelm II for a diplomatic solution. Instead, he was forced to resign. The kaiser was a man of unimpeachable mediocrity, and he still believed that the war could be won by military might alone. We all know how that worked out.

THE SECRET TO WINNING

Often people use the word "peace" when they mean "victory." Half of winning is knowing what it looks like. Western countries have forgotten, judging by their war record since 1945. Their assumptions about victory are still moored to the conventional war theories developed by Clausewitz and others in the eighteenth and nineteenth centuries. Now it is the twenty-first century, and no one wins that way anymore.

War is more than the military clash of wills. Kühlmann knew this but was ignored. He understood war's most important secret: war is armed politics. This is the only true law of war. War is not inherently evil, and it is not even inherently military. It does involve organized violence, or the threat of it, and it does wreak human suffering. Naive, bloodless solutions to bloody problems fail, resulting in greater bloodshed. The greatest strategic thinkers in history all agree on this, from Sun Tzu to Clausewitz.

Militaries misread this simple but powerful truth, preferring to blow away the enemy first, then negotiate political issues second. This is wrong. Typically, this approach does not end wars; it prolongs them. Doing so means you may defang a threat for a while, but you won't have solved the underlying political problems that sparked the given conflict in the first place. The enemy will rise again, and you will have to subdue it again, in an endless cycle of violence. This is why the Romans besieged the impossible fortress of Masada: to exterminate the insurrection then and there.

"War is armed politics" means victory is as much political as military. In other words, you do not need to win battles to win the war. This may sound obvious, but the Western great powers remain fixated on battlefield victory despite the bitter lessons of the last seventy years. The United States is a supreme example. After the initial invasion of Iraq, President George W. Bush stood on the deck of the aircraft carrier USS *Abraham Lincoln*, declaring victory with a large "Mission Accomplished" banner behind him, flapping in the wind. The US military had swept aside Saddam Hussein's army, a decisive battlefield victory for America. Or so it thought. The Iraq War would endure for another eight years, and the United States remains embroiled in that country today and potentially for years to come. "Mission Accomplished" has devolved into a meme denoting clueless failure.

Nothing has changed since then. The United States and its allies have been unable to "win the peace" (whatever that means) in Afghanistan, Iraq, Syria, and everywhere else they fight. Tactically, Western militaries remain unrivaled. Any country that engages the US military in open battle will be crushed, should America choose. However, Western militaries are obsolete strategically and easily frustrated by weaker foes. There are easier ways to win than open warfare, and such strategies do not require a big military, or even a military at all. Firepower is not needed to win war. This is how David defeats Goliath.

BUT DO THE WEAK WIN?

On January 30, 1968, tens of thousands of North Vietnamese troops launched the Tet Offensive. A truce was supposed to have been in place to mark Tet, the lunar new year holiday, but instead the North Vietnamese began a massive attack into every area of South Vietnam, including Saigon. Until then, the city had been considered unassailable.

A suicide squad stormed into the US embassy's compound in Saigon. Chuck Searcy was a twenty-year-old enlisted soldier at the time.

"After midnight, the siren went off—the alert siren—which was our signal to go to our posts," Searcy recalled. "So everybody gets out of the bunks, grumbling and bumbling, and put on all our gear and went out to the perimeter, assuming that 15 minutes later, we'd have the all-clear signal; we'd go back to sleep. But then a captain came around the perimeter in a jeep with a loudspeaker announcing that this was not a practice alert, that Ton Son Nhut Air Base had been overrun and Saigon was getting hit very hard."

Searcy was with the 519th Military Intelligence Battalion, and he says the attack caught almost everyone by surprise. In addition to Saigon, the Viet Cong and the North Vietnamese Army struck thirty provincial capitals and the old imperial capital of Hue.

American and South Vietnamese forces responded quickly, recapturing lost ground and decimating an enemy that had "finally come out to fight in the open." Communist losses were heavy, and they were finished as a fighting force. It was one of the most resounding defeats in military history—until it became a victory.

News footage showed the fighting in Saigon and Hue. The Tet Offensive shocked Americans at home, who thought the war was nearing victory. The White House's earlier optimistic assessments were proved to be hokum overnight, creating a credibility gap that widened into a chasm as the fighting got worse. On February 18, the military posted its highest casualty numbers in a single week for the entire war: 543 killed and 2,547 wounded. A few days later, the US government announced a new draft calling for

48,000 men, the second-highest of the war, and President Johnson considered mobilizing an additional 50,000 troops from the reserves. A week later, Robert McNamara, the secretary of defense who had overseen the escalation of the war before turning against it, resigned or was fired.

Americans were alarmed. Walter Cronkite, the anchor of *CBS Evening News*, returned to Vietnam to see for himself what was happening. He had been a war correspondent during World War II and had reported from Vietnam during America's early involvement. In 1972, a poll would determine that he was "the most trusted man in America."

What Cronkite saw was disheartening. In a special TV broadcast, Cronkite did something that changed Americans' perception of the Vietnam War. "We are mired in stalemate," he said, and "the only rational way out then will be to negotiate, not as victors, but as an honorable people who lived up to their pledge to defend democracy, and did the best they could."

President Johnson, watching the broadcast, said, "If we've lost Walter Cronkite, we've lost the country." A few months later, Johnson announced that he would not run for reelection. He also said there would be a pause in the air attacks on North Vietnam as "the first step to deescalate" and promised America would substantially reduce "the present level of hostilities." The United States was on the way to leaving Vietnam.

This was the moment that the North Vietnamese won the war, not by firepower but with the nightly news. They knew they could never defeat the United States military in open battle, so they generally avoided it. The Tet Offensive was a military failure but a propaganda success, dealing a strategic deathblow to the American war effort. Unlike the American leadership—which was stuck in a conventional war mind-set—the North Vietnamese leaders understood that waging war in the television age depends as much on propaganda as it does on success in the field. This is especially true when fighting democracies, because their citizens can hire or fire policy makers.

The United States won the tactical fight on the ground, while the North Vietnamese won the strategic one on the American home front. From the beginning, the North Vietnamese ran a covert campaign of strategic influence against the United States on the other side of the globe. Their logic was flawless: if you cannot defeat the US military in battle, then persuade its boss, the American people, to withdraw the troops.

Hanoi used secret agents to manipulate the American press pool in Saigon by acting as trusted sources who fed misinformation. Some even became major journalists. Pham Xuan An was a reporter for Reuters and *Time* magazine, and also a frequent unnamed source for major outlets like the *New York Times*. After the war, he was revealed to have been a colonel in the North Vietnamese Army the whole time.

Hanoi also wooed American influencers, such as the movie starlet Jane Fonda and Ramsey Clark, a former attorney general. Fonda appeared at a press conference wearing a red Vietnamese dress, saying she was ashamed of American actions in the war and would struggle to end the fighting. In another photo, she posed with North Vietnamese soldiers in Hanoi, smiling and sitting in the gunner's seat of an antiaircraft gun used to shoot down US aircraft. American soldiers scorned her as "Hanoi Jane," and some on Capitol Hill accused her of treason. But by then it was too late, and the North Vietnamese influence strategy was paying off.

Americans began questioning the war. Negative media coverage helped fuel antiwar protests across the country, and the Tet Offensive was timed to coincide with a contentious presidential election that would decide the fate of US involvement in the war. Hanoi worked the Americans at home while the US military worked the battlefield. "It was essential to our strategy," said Bui Tin, a former colonel in the North Vietnamese Army. He said the Vietnamese Politburo was keenly aware of US domestic politics, and its members would listen to world news at nine a.m. every day on the radio to follow the growth of the American antiwar movement. Meanwhile, leaders in Washington dithered to the point of dereliction. For the United States, the Vietnam War was not lost in Vietnam—it was lost at home.

This is how David beats Goliath. In the case of Vietnam, information proved more decisive than firepower, allowing the North Vietnamese to achieve their war objectives while denying the Americans theirs. This is victory. The North Vietnamese did not need a big military or superior technology; instead, they used nonkinetic means like the media to portray the war as "mired in stalemate," which weakened American resolve to the point of withdrawal. Weaponizing information is effective because it controls the narrative of the conflict, asking the question of why people should fight and die (or not).

After the Vietnam War, an American colonel turned to his North Vietnamese counterpart and said, "You know you never defeated us on the battlefield." The North Vietnamese colonel pondered this a moment. "That may be so," he replied, "but it is also irrelevant."

The North Vietnamese were not the first to win by weaponizing information. Benjamin Franklin was a master propagandist during the American Revolution. To garner support for the plucky revolutionaries among the British masses, Franklin published fake news in a counterfeit issue of the *Boston Independent Chronicle*, an influential newspaper. Writing all the articles himself, he had the newspaper discreetly put into the hands of British tabloid editors, who reprinted the articles. One depicted wartime atrocities by Indians at the behest of the British. Specifically, it described the delivery of "eight Packs of Scalps, cured, dried, hooped and painted, with all the Indian triumphal Marks." The boxes, Franklin's article claimed, contained hundreds of American scalps, including: "193 Boys' Scalps of various Ages," "211 Girls Scalps, big and little," and "29 little infants Scalps of various sizes. . . . A little black Knife in the middle to shew they were ript out of their Mothers' Bellies." Franklin's "scalping letter" went viral, and the hoax was reprinted in England, Amsterdam, and the colonies. Readers were aghast at the British government, helping to create the conditions that led to the colonials' victory.

T. E. Lawrence believed that "the printing press is the greatest weapon in the armory of the modern commander." Lawrence operated in a newspaper age; the North Vietnamese fought in a television age; and we are said to live in an information age. So what of the future? The truth is, a clever person can weaponize almost anything: refugees, information, election cycles, money, the law. Power is mutable in war, and victory belongs to the smart rather than the strong. "Irregular war [is] far more intellectual than a bayonet charge" was one of T. E. Lawrence's favorite quips. General Vo Nguyen Giap, the commander of the North Vietnamese military, read a lot of Lawrence, and here is how he explained his victory:

> The American soldiers were brave, but courage is not enough. David did not kill Goliath just because he was brave. He looked up at Goliath and realized that if he fought Goliath's way with a sword, Goliath would kill him. But if he picked up a rock and put it in his sling, he could hit Goliath in the head and knock Goliath down and kill him. David used his mind when he fought Goliath. So did we Vietnamese when we had to fight the Americans.

CHOOSE YOUR WEAPON OF WAR

Victory belongs to the cunning, not to the strong. To the conventional warrior, there is only firepower and annihilation. Victory is a zero-sum calculus of who's left standing. You can still win this way, but it's bloody, hard, and increasingly rare. More artful ways of winning include denying the enemy its victory conditions. Or simply refusing to die, as George Washington did with the scrappy Continental Army in 1781. "The guerrilla wins if he does not lose. The conventional army loses if it does not win," explained Henry Kissinger during the Vietnam War.

Clever strategies exist that allow the weak to reliably defeat the strong. Weaponizing time is one, and it works. It was used by the Roman general Fabius against Hannibal's superior armies in the Second Punic War (218–201 BCE), and known in the Middle Ages as the Fabian strategy. George Washington employed it against the redcoats during the American Revolution. The Russians defeated Napoleon's massive invasion of 1812 with it, the biggest upset in military history. T. E. Lawrence applied it to help the Arabs overtake the Ottomans, and Mao Zedong utilized it as he fought the nationalists, then the Japanese, and then the nationalists again—always outgunned, outmanned, and outmaneuvered—until he won. Today, the Taliban, al-Qaeda, and just about everyone else uses it. Curiously, conventional militaries rarely study it.

Here's how the strategy works. The weak can defeat the strong, if certain preconditions exist.

IF THE STRONG ARE

- a big military that fights using conventional war strategies and tactics;
- a foreign invader that most of the local population dislikes;
- spread too thin on the ground;
- unable to learn well and adapt slowly; and
- fighting a war of choice, not a war of survival.

IF THE WEAK ARE

- fighting a war of survival, not a war of choice;
- locals defending their homeland from an outside threat;
- unified in a cause, ideology, or identity;
- able to access a sympathetic population (it need not be actively friendly); and
- prepared to bleed.

THEN THE WEAK DEFEAT THE STRONG IF THEY

- avoid battle and focus on survival;
- gain enough popular support for safe havens, resupply, intelligence, and fighters (this support can be earned or coerced);
- mobilize the population into guerrilla units, bound by a common cause, ideology, or identity;
- wage a guerilla war campaign with hit-and-run tactics, and live off supplies captured;
- gain local and international sympathy by provoking an over-reaction by the strong against the weak;
- turn the population against the strong by showing they are not so strong, through propaganda, sabotage, assassination, guerilla war, and terrorism;
- draw the enemy deep into the country's interior, where the guerrillas can bleed them dry;
- protract the conflict: the longer the war drags on, the costlier it becomes for the strong in terms of troops, resources, and political will back home; and
- make it too costly for the strong to remain, after which they will eventually leave of their own accord.

This strategy succeeds because it is cheap to carry out and expensive to suppress. It does not require a big military, or a military at all, and the more conventional the Goliath is, the worse it gets for him. Given enough time, the weak win, because the strong collapse under the weight of their own occupation, and they leave voluntarily.

THE MYTH OF BIFURCATED VICTORY

Every year, my American students at the war college get into furious debates over who won the wars in Vietnam, Iraq, and Afghanistan. The arguments in each case sound remarkably similar, even though almost fifty years separate these conflicts. The students are high-ranking officers and are likely to be going higher, but when this topic comes up, the rank comes off.

"I'm not sure who won Iraq," said an army colonel one year, a veteran of three combat tours. "All I know is that we won every battle, and that's 'not losing.' Maybe we didn't win, but we sure as hell didn't lose."

"Who cares if we won every battle?" said a diplomat from the State Department. "We didn't achieve any of our war aims, and that sounds like losing to me. You forget that the enemy gets a vote, too."

"Maybe we didn't win the war in Iraq," said a marine colonel and combat veteran. "But we didn't lose, either. We abandoned the war, not the same as losing."

Several nodded in agreement, except one. An Iraqi army colonel just shook his head, perhaps knowing that ISIS was overrunning his country as we were having this discussion.

Even now, the easiest way to get into an argument at a VFW bar is to mention Vietnam. Seared into all who fought it—and many who merely lived through it—that conflict remains a bitter stew of second-guessing and recriminations. Few will admit the United States lost despite overwhelming evidence to the contrary. It is bitterly hard to accept that friends died and that our country didn't achieve what it set out to do.

Anytime David wins, Goliath's explanation is always the same, whether it's Vietnam, Iraq, or Afghanistan: We didn't win, but we didn't lose. General William Westmoreland, the commander of US forces in Vietnam, summed it up for a generation: "Militarily, we succeeded in Vietnam. We won every engagement we were involved in out there." The US military says the same thing today about Iraq and Afghanistan.

This is the myth of bifurcated victory: that one can win militarily, yet lose the war. But this is like the old joke: "Doctor, the operation was a success, but the patient is dead." Victory and defeat have meaning only in political terms. Failure to translate military victories into political ones equals defeat, and this is how big militaries lost in Vietnam, Iraq, and Afghanistan.

THE "TACTIZATION" OF STRATEGY

The "tactization" of strategy is another reason the West struggles in modern war. As Sun Tzu is said to have counseled, "Strategy without tactics is the slowest route to victory. Tactics without strategy is the noise before defeat." Let me explain.

War has three levels: tactical, operational, and strategic. Tactical is the base level of war, and it's the realm of the soldier. Tactics are about maneuvering small units on a battlefield, and conducting airstrikes and individual actions at sea. Military historians often focus on the tactical level, concentrating on great battles rather than the societies that produced them. Hollywood also glorifies the tactical level, from John Wayne in *Sands of Iwo Jima* to Tom Hanks in *Saving Private Ryan*.

Next is the operational level. It involves integrating several military campaigns within a theater of war. During World War II, in the European theater, the Allied landing on the beaches of Normandy and the push to Berlin is a good example. Military planners had to coordinate multiple campaigns, from that of Patton's Third Army to Montgomery's Twenty-First Army Group, and manage the outcome of hundreds of tactical encounters. The military likes to call this feat "operational art."

At the top is strategy. War at the strategic level extends beyond military means, and it encompasses all instruments of national power—economic, diplomatic, social, political, information, military—to achieve a state's national interests. Done well, each country develops a blueprint or grand strategy for accomplishing this, like a theory of victory. The strategic level is also where war is most political. How do you know you're functioning at the strategic level? When military victories translate into political ones and vice versa—the military-political nexus.

Setbacks at the tactical and operational levels can be overcome, but failure at the strategic level can cause the whole house of war to collapse. This is why winning can be realized only at the strategic level, and the only legitimate measure of victory is this: Did the war attain the goals it set out to achieve? Anything beyond the scope of that question does not make sense as victory, and simply declaring victory and leaving the conflict fools no one, especially history.

Many today—some at the highest levels—confuse tactics with strategy, focusing on combat overmatch on the ground without linking it to a war's political objectives. Tactical dominance achieves little in modern or future wars, as has been seen from

Vietnam to Afghanistan. Precision drone strikes kill enemy leaders and other high-value targets, yet the threat grows. For every terrorist killed, three more appear. Troops joke that this is the "Whac-A-Mole" strategy, named after an arcade game and used colloquially to denote a repetitious and futile task: each time the enemy is "whacked," it pops up again somewhere else. Whac-A-Mole fails because it is a purely tactical approach to a strategic problem. Tactical encounters alone rarely alter political situations.

One solution to this and the bifurcated victory problem is improved strategic education. Big militaries are full of tactical thinkers rather than strategic ones because we educate our military this way. From its service academies like West Point to its war colleges, the United States has the finest professional military education system in the world. However, this system often confuses tactics with strategy. Many study the Battle of Gettysburg in their strategy classes. Gettysburg was a tactical event, not a strategic one. Curriculums fixate on nineteenth-century war theorists such as Carl von Clausewitz and Alfred Thayer Mahan, fathers of conventional war who valued tactics over strategy. Antoine-Henri Jomini, the most influential war theorist of the nineteenth century, exalted this approach as "grand tactics." For these thinkers, battles like Waterloo win wars, and wars like Vietnam would be incomprehensible. These theorists are well past their expiration date, since no one fights like it's 1830 anymore.

The quality of strategic education is worse at civilian universities. Most of them do not teach strategic studies at all, a significant oversight, and those that do typically offer abstruse academic theories that amount to magical thinking. There are exceptions, of course. Every university and war college has its share of brilliant thinkers, and even better teachers. In the field of strategic studies, they are often scholar-practitioners who have read enough theory and gained enough field experience to have glimpsed war's future.

Strategic education starts too late in the US military. Military academies like West Point (think undergraduates) barely teach it, and war colleges (think graduate school) teach it too late. The average war college student is a senior officer with fifteen years of service behind him or her. Leaders must learn to think strategically as cadets, not as colonels. It's often too late by then. This is because tactics and strategy require two different kinds of thinking, and they are diametrically opposed. One is complicated while the other is complex. Imagine a Boeing 747 versus Congress. The 747 is one of the most complicated machines on the planet. It has a zillion parts, but with enough mental horsepower it can be taken apart, reassembled, and flown. In other words, it's solvable. Not so with Congress. It is composed of 535 people who act independently and often unpredictably. Congress cannot be solved like a big puzzle, which makes it complex. In systems theory, complicated systems can be solved, while complex ones cannot.

The same is true for war—tactics are complicated, but strategy is complex. A big part of strategic education involves getting officers to think about strategic complexity when they have spent fifteen years thinking about tactical complicatedness. Most officers have degrees in engineering, a discipline structured around devising solutions to complicated problems. All the military academies have hard-core engineering curriculums. Every English major at Annapolis receives a Bachelor of Science, earning more credit hours in the sciences than in the arts. Such an approach ill equips young leaders to become strategic thinkers.

A recent concept called "hybrid war" tries to bridge the chasm between tactical versus strategic thinking, and the related problem of conventional versus unconventional warfare. The idea comes from Frank Hoffman, a retired marine colonel turned war scholar. Hybrid war describes modern conflicts as a mixture of conventional and unconventional warfare at all three levels of war. For example, Russia took Crimea using conventional hardware like tanks and unconventional assets like little green men. This powerful idea helps conventional warriors make the leap to the postconventional war world. It gives them enough familiarity to grasp today's strategic complexities without blowing their tactical circuitry. The theory of hybrid war is important, but it does not go far enough.

The West needs to relearn how to win strategically. Strategy is art while tactics is science, so Western militaries need people who can think about war creatively. This will require a new kind of strategist.

DEVELOP WAR ARTISTS

It's 2016, and a tall, lank marine walks across the stage to a podium. Washington, DC, swelters in June and we are roasting outside, but you would never know it from his crisp jacket and tie. It's as if the man doesn't sweat. Across his left breast are rows of medals. Each epaulet has four shiny stars. General Joseph Dunford is chairman of the Joint Chiefs of Staff and the United States' highest-ranking member of the armed forces. He is also our graduation speaker at the National Defense University, which is sometimes called the "chairman's university," since that position oversees it.

"You know, I'm mindful that commencement ceremonies are not about long speeches," he says, and we all chuckle with optimism.

Five thousand of us sit under a palatial tent, melting in ninety-degree heat and rainforest humidity. The military is buttoned up in Class A uniforms, and civilians wear wool suits. I sit with the faculty garbed in full-length academic gowns meant for cooler climates, not the fever swamp of Washington, DC.

"How many of you are going to the Joint Staff?" Dunford jokes, referring to officers assigned to the Pentagon after graduation. Hands go up. "Your vehicles should actually be running in the parking lot because, you know, since you're already here, we got the welcome-aboard set up for you at 1700 this afternoon."

Laughter is faint. A general once told me that duty in the Pentagon is like the naval battle in *Ben-Hur*: "Row well, and live."

Jokes dispensed, Dunford talks about the changing profession of arms. It's clear that he's not talking to military members alone but to all of us, as national security now requires more than a big military. War colleges now have senior students from across the federal government and foreign countries.

"As most of you know, 2016 marks one hundred years since the Battle of Verdun, a battle that claimed over five hundred thousand lives. Verdun provides a pretty good case study, in my estimation, for change," he says, then discusses how tens of millions of lives were lost needlessly in the world wars because decision makers on all sides were too slow to adapt to new forms of warfare. They had little excuse. Thinkers between the wars like Billy Mitchell and J.F.C. Fuller anticipated some of these problems, but they were ignored by the establishment. Then Dunford turns to the present.

"We're already—To be honest with you, we're already behind," he concludes. "We're already behind in adapting to the changed character of war today in so many ways."

This verdict should worry us all, coming from the United States' top general. America and other Western nations are not struggling in modern wars because they lack good troops, training, or equipment. In each area, they are the best. The reason is because Western nations are not yet adapting to the new character of war, and, according to Dunford, "not thinking out to the future in an innovative way."

General Dunford's main point to the graduating class is this: We've been here before. Militaries are slow to evolve their thinking about warfare, and it gets soldiers killed; therefore, we have a moral obligation to do better. Yet history shows us that few do this, or can.

In the future, we will need more than warriors—we will need war artists. War is more like jazz than engineering, and we need strategists who can think this way. How does one develop war artists? Start with the liberal arts; learning *how* to think is more important than knowing *what* to think. Military academies and other commissioning sources need to encourage the arts and expand their repertoire to entice students to discover their inner strategist. War colleges need to take their strategic education out of the nineteenth century and include non-Western traditions.

Strategic education must start young. Why do we wait fifteen years before we introduce strategy into an officer's intellectual diet? By then, the tacticians will have risen to the top, and many of the strategists will have left the service in frustration. Civilian war artists face an even steeper climb. The strategic education offered at civilian universities is appalling, often taught by people who have never ventured beyond the ivory tower, much less into war. We possess the talent but are not cultivating it.

We need to identify strategic geniuses early and develop them in a separate program, one that's open to military and nonmilitary personnel alike. Why do we assume only military leaders make effective war strategists? Lincoln and Churchill were brilliant. Not everyone can be a great strategist, but an effective, creative strategist can come from anywhere.

Orson Scott Card's science fiction novel *Ender's Game* lays out a thought-provoking example. Military strategists in his book are identified when they're children based on their intellectual and creative abilities, as well as by their moral ones. Of course, for those who know the book, the warfare for which the young strategists are prepared is starkly conventional. Science fiction is often a mirror to the present. But the image of an academy for budding strategists is compelling. What skills and character traits would we prize? What curriculum would we build? What mentorship would such students need? What experiences must they have? These are the questions we should be asking so we can spot and recruit a cadre of strategists who will innovate new ways of war.

"All things are ready, if our mind be so," declared Shakespeare's *Henry V* before the Battle of Agincourt. The English won. An agile strategic mind is more important than smart bombs, gee-whiz technology, or numerical superiority. None of these things win war without a quality strategy behind them. War artists can win the future, if we cultivate them. The purpose of studying war is to reduce it, or wage it as efficiently as possible when there is no other choice. This is important if it means saving lives and countries.

WINNING THE FUTURE

One July morning in 2006, two Israeli Humvees patrolled their country's northern border with Lebanon. Somewhere on the other side was Hezbollah, the feared terrorist group and Iranian proxy militia committed to the destruction of Israel. It owned southern Lebanon, regardless of what people in Beirut thought.

The patrol started like any other. The vehicles skirted the fortified border fence, winding around hills and valleys. Birds were the only sound. Then Katuysha rockets ripped across the sky, launched from the Lebanese side of the border.

Hezbollah, the soldiers thought. The rockets hit a village not far away, the ground shaking on impact.

"*Kadeema!*" shouted the squad leader. "Let's go!"

The two Humvees sped toward the village, hoping to aid survivors, when they were ambushed. The rockets were merely a diversion to lure them into a death trap. A vicious firefight ensued. When the smoke cleared, three Israeli soldiers were dead, another five lost in a failed rescue attempt, and two had been taken prisoner by Hezbollah.

Hours later, the Israeli prime minister, Ehud Olmert, ordered the military to attack Hezbollah in Lebanon. Beirut was a bystander.

Days later, Olmert spoke with venom against Hezbollah in front of the Knesset, Israel's national legislature. For years Hezbollah had been launching attacks from Lebanon into Israel, and for years Israelis had been dying. Worse, the Lebanese government had taken no interest or responsibility for the situation.

"No more!" Olmert shouted. He promised the war would be quick and easy because the terrorist group was no match for the Israel Defense Forces, the most powerful and technologically advanced military in the Middle East. The war would achieve four things, he said: regain the two kidnapped soldiers, achieve a complete ceasefire, force Hezbollah from southern Lebanon, and ensure that the Lebanese army kept the group out. Olmert ended with, "We will triumph!"

When Hassan Nasrallah, the leader of Hezbollah, heard this declaration, he must have become giddy. Israel's military possessed complete combat overmatch over Hezbollah, but what did it matter? Olmert had just announced four unrealistic war objectives, and—even more damning—he'd done it on international TV. There was no going back for Olmert. If Hezbollah could prevent just one of Israel's victory conditions, Israel could not claim victory. In war, as in life, you must do what you say to maintain credibility.

Over the next thirty-four days, Israel launched a conventional war that pounded southern Lebanon into sand. Whole sections of Beirut were flattened, and the international airport was also bombed. All told, seventy bridges, ninety-four roads, three airports, four ports, and nine power plants were destroyed. The country was blitzed. By the war's end, ten times more Lebanese had been killed than Israelis, and Israel controlled southern Lebanon.

Yet Israel lost. In a now-familiar refrain of modern war, Israel won every battle but lost the war—and to a weaker enemy. Hezbollah won, not by conquering Israel, but by denying Israel its victory conditions. Olmert was strategically stupid when he promised that Israel would recapture its two soldiers. Who can't hide two people? As long as Hezbollah held the captives, Israel could never declare victory. Israel finally recovered the soldiers' bodies—two years later.

Israel lost because it set victory conditions that were unattainable through military force, then launched a military campaign to attain them. Worse, Olmert publicly announced these victory conditions, giving him zero plausible deniability in case things went badly, and they did. Winning for Hezbollah was easy. All it had to do was let Israel fail.

The Israel-Hezbollah conflict illustrates war's new rules. If it had been an old-fashioned conventional war, Israel would have won. It captured more land, killed more people, destroyed more critical infrastructure, and flew its flag over the enemy's territory. But it didn't win, because conventional war is dead, rendering Israel's superior firepower and combat technology irrelevant (rules 1, 2, 5). The idea of war versus peace does not apply to southern Lebanon or any other place on Israel's border (rule 3). Civilians were targeted, and their deaths outnumbered combatant casualties (rule 4). The fighting took place in another state, Lebanon, but its military was curiously AWOL because the conflict was between Israel and a nonstate actor (rule 7).

Ultimately, Israel succumbed to Iran's shadow war (rule 9). Hezbollah is an Iranian proxy militia, although it's not a Tehran sock puppet and exercises a fair degree of autonomy. Israel's strategic missteps allowed its enemy to determine victory (rule 10). Hezbollah has since redeployed itself to Syria, a disorderly battleground replete with clashing nonstate actors and mercenaries (rules 6 and 8). Syria shows us the future of war, in contrast to the cockamamie visions of Western futurists.

Israel lost because it did not understand the new rules of war, but it is also a model because it has a learning military. After the war, the country engaged in strategic soul-searching. A special commission, headed by former justice Eliyahu Winograd, was set up to investigate the failed military campaign. Recriminations spread through the Israel Defense Forces and society more generally. Out of this furnace a new strategy was forged, one that recognizes the changing character of war. Nearly ten years later, the Israel Defense Forces unveiled a strategy that includes elements of shadow war, something it calls the Campaign between Wars. Israel is leaving conventional war behind and is stepping into the future, joining Russia and China. The West should follow in its footsteps.

FUTURE WAR

The West will be defeated if it cannot adapt to the future of war. This is inevitable. Western powers are already losing on the margins to threats like Russia, China, and others that have made the leap forward and grow bolder each year. Eventually someone will test us and win.

The West has forgotten how to win wars because of their own

strategic atrophy. Judging by how much money the United States invests in conventional weapons like the F-35, many in our country still believe that future interstate wars will be fought conventionally. But although Russia and China still buy conventional weapons, they use them in unconventional ways, as was revealed in Ukraine and the South China Sea. Russia even cut its military budget by a whopping 20 percent in 2017, yet it shows no sign of curbing its global ambitions. Its leaders understand that war has moved beyond lethality.

Conventional war thinking is killing us. From Syria to Acapulco, no one fights that way anymore. The old rules of war are defunct because warfare has changed, and the West has been left behind. War is coming. Conflict's trip wires are everywhere: black market nukes that can melt cities; Russia taking something it shouldn't and NATO responding in force; India and Pakistan nuking it out over Kashmir; North Korea shelling Seoul; Europe fighting an urban insurgency against Islamic terrorists; the Middle East goes nuclear; or the United States fighting China to prevent it from becoming a rival superpower.

As bad as these threats sound, they are episodic and hardly the worst. Festering systemic threats like durable disorder will shake global security in the twenty-first century, as evidenced by the increased number of armed conflicts in our lifetime. Traditionalists who view war purely as a military clash of wills will be doomed, no matter how big their armed forces, because they do not comprehend war's political nature, while their enemies do. There are many ways to win, and not all of them require large militaries.

Changing the way we fight means forging new instruments of

national power, starting with how we think. The first step is jettisoning what we think we know about war. Our knowledge is obsolete. The second step is understanding the art of war for the coming age, so that we may master it rather than be mastered by it.

In the future, wars will move further into the shadows. In the information age, anonymity is the weapon of choice. Strategic subversion will win wars, not battlefield victory. Conventional military forces will be replaced by masked ones that offer plausible deniability, and nonkinetic weapons like deception and influence will prove decisive. Shadow war is attractive to anyone who wants to wage war without consequences, and that's everyone. That is why it will grow.

Future wars will not begin and end; instead, they will hibernate and smolder. Occasionally they will explode. This trend is already emerging, as can be seen by the increasing number of "neither war, nor peace" situations and "forever wars" around the world. The World Bank's *World Development Report 2011* found that never-ending violence is on the rise, despite all the peace efforts around the world. Social science research confirms this, showing that half of all negotiated peace settlements fail within five years. "War termination" is already an oxymoron. Expect this trend to grow.

Mercenaries will once again roam battlefields, breeding war as their profit motive dictates. International law cannot stop them, while the demand for their services rises each year. Things once thought to be inherently governmental are now available in the marketplace, from special forces teams to attack helicopters. This is one of the most dangerous trends of our time, yet it's invisible to most observers. That's by design. Private warfare is the norm in military history, and the last few centuries have been anomalous.

When money can buy firepower, then the super-rich will become a new kind of superpower, and this will change everything. As states retreat, the vacuum of authority has bred a new class of world powers, from multinational corporations to superwarlords to billionaires. Now these powers can rent private armies, so expect wars without states. This trend will grow, fueled by a free market for force that generates war but cannot regulate it. Today's militaries have forgotten how to fight private wars, leaving us all exposed.

To the conventional warrior, this all looks like disorder and instills panic. The world is burning without a way to put the fire out. But the new warrior sees something different. States are dying as a concept and are being replaced by other actors, who also fight. How they fight is not disorder—it's the future of war. Rather than panic, let's master this future.

The good news is that we can win in an age of durable disorder if we understand the new rules. This begins by transforming militaries from conventional forces to postconventional ones, and by upgrading our strategic education. We should invest in people rather than machines, since cunning triumphs over brute force, and since technology is no longer decisive on the battlefield. We also need a new breed of strategist—I call them war artists—to contend with new forms of conflict, such as private war.

Half of winning is knowing what it looks like, and this requires a grand strategy. In an age of durable disorder, our grand strategy should seek to prevent problems from becoming crises and crises from becoming conflicts. Attempting to reverse disorder is a Sisyphean task because such disorder is the natural condition of world affairs—again, it's the recent centuries that have been abnormal.

Handling persistent disorder is like undertaking chronic pain management for an illness rather than attempting to find a cure. Also, no one should weep for Westphalia. The bloodiest wars in history—the world wars—occurred on its watch.

War is going underground, and the West must follow by developing its own version of shadow warfare. Special operations forces should be expanded, since they can fight in these conditions, and the rest of the military needs to become more "special," too. The West must do a better job at leveraging proxy forces and mercenaries. But the true weapon of choice will be the foreign legion. It will combine the punch of special operation forces with the staying power of a normal military unit, all without the problems of proxy militias or mercenaries.

In the future, victory will be won and lost in the information space, not on the physical battlefield. It's absurd that the West has lost information superiority in modern war, given the heaps of talent in Hollywood, on Madison Avenue, and in London. The West's squeamishness about using strategic subversion only helps its enemies. Sun Tzu and the Thirty-Six Stratagems for War are a good place for us to start overcoming this squeamishness. Let a war artist take it from there. The West can win if it fights with the New Rules of War. Only then will we be secure.

THE CHOICE BEFORE US

"No bastard ever won a war by dying for his country. You won it by making the other poor dumb bastard die for his country."

When General Patton spoke these words seventy-five years ago, they were true. His troops were about to embark on the greatest amphibious assault in history: D-day. Over the course of the "longest day," 160,000 Allied troops seized a slab of beachhead in Normandy, France. More than 10,000 of them died that day, and individual acts of valor turned a potential military catastrophe into a triumph. From there, the Allies began the long march to Berlin, ending the Nazi empire.

Patton's words are no longer true.

Today, bastards do not die for their country; they die for their religion, their ethnic group, their clan, money, or war itself. A few, like Afghans and Somalis, say they fight for their country, but the "country" in question is a metaphor and not a modern state. In fact, were there a functional state in those situations, they would probably fight that, too. Patton, were he alive, would be holding his head in his hands.

Countries need to evolve the way they fight, but can they do it? History teaches us that this transformation is difficult. Billy Mitchell was court-martialed in 1925 for having the audacity to suggest that the future of war would be dominated by airplanes and aircraft carriers, not by battleships. He predicted Pearl Harbor sixteen years before it happened. His superiors laughed as they convicted Mitchell because it was easier than listening to him—only to be caught "by surprise" on December 7, 1941.

Strategic dogma is stubborn because everything about it is existential. If you get it wrong, the nation dies. This is why strategic leaders are leery to experiment with new approaches, and perhaps why the military calls its tactical playbooks "doctrine." But such devotion also gets people killed. Typically, blood is required—a huge amount of it—before nations change their way of war, and sometimes not even then.

World War I is a good reminder. Strategists on all sides were stuck in their own past glory days: Napoleonic warfare. However, by the time World War I broke out, fighting had moved well beyond that of Napoleon's day, and millions died pointlessly because leaders had no strategic imagination. Or they just toed the line. Politicians commanded the generals to win, and they in turn ordered waves of soldiers to assault fortified trenches, only to see them slaughtered by machine guns. Still, this didn't stop the generals from doing the same thing the next morning. During the Battle of the Somme, the British suffered sixty thousand casualties in one week. That's more than all the Americans killed in the Vietnam War. The Battle of the Somme was a meat grinder, stealing 1.2 million lives and achieving nothing.

History is replete with Sommes because it is the nature of militaries to resist change. Many generals are rigid in their understanding of war, how it should be waged, and how it should be won. This hardness of mind is necessary to wage war, but not to think about and plan for it. Dogmatic approaches can lead to terrible endings. World War I suffered nearly forty million dead and wounded, and for what? A Pyrrhic victory for the Allies that midwifed World War II. Mitchell was right: nations must learn to fight anew, even if they do not want to.

If there is anything to learn from military history, it's this: warfare evolves before fighters do. War in our time has already changed, but most nations have not. This includes their militaries, political leaders, intelligence agencies, national security experts, media, academic institutions, think tanks, and members of civil society who care about armed conflict. The West's way of war has evolved little since Patton's day, and this rigidity has cost us needless lives, just like at the Somme.

There is a choice before us. Either we spill enough blood in battle until we finally realize our problem, or we choose to change now. No one ought to select the former, but the latter is difficult. It will require disruptive thinking and bold steps that conventional warriors will reject but troops on the ground will understand.

It will not be easy, but as any soldier will tell you, nothing worth fighting for is.

ANNEX: THE THIRTY-SIX ANCIENT CHINESE STRATAGEMS FOR WAR

"All war is deception." These are the immortal words of Sun Tzu, the ancient Chinese general and war theorist. He wrote *The Art of War* in the fifth century BCE, and it has never stopped being read. The Thirty-Six Stratagems, written by an unknown author, have also been passed down from general to general since Chinese antiquity. The following table presents an overview. Conventional warriors will reject these ideas but those of the future will not, and China uses them today. War is timeless, as are these stratagems.

	The Original 36 Stratagems	Contemporary Maxims
1	Fool the Emperor and Cross the Sea	Act in the open, but hide your true intentions.
2	Besiege Wei to Rescue Zhao	Attack the enemy's vulnerable area.
3	Kill with a Borrowed Knife	Attack using the strength of an ally.
4	Await the Exhausted Enemy at Your Ease	Exercise patience and wear the enemy down.
5	Loot a Burning House	Hit the enemy when he is down.
6	Clamor in the East while attacking in the West	Fake to the right; attack to the left.
7	Create Something from Nothing	Turn something that is not substantial into reality.

	The Original 36 Stratagems	**Contemporary Maxims**
8	Pretend to Take One Path While Sneaking Down Another	Pretend to care about an issue and later give it up to get what you really want.
9	Watch the Fire Burning from Across the River	Allow one enemy to fight another while you rest and observe. Later, defeat the exhausted survivor.
10	Hide Your Dagger behind a Smile	Befriend your enemy to get his defenses down, then attack his weakest point.
11	Sacrifice a Plum Tree to Save a Peach Tree	Trade up. Take a small loss for a large gain.
12	Take the Opportunity to Pilfer a Goat	Take advantage of every small opportunity.
13	Beat the Grass to Startle the Snake	Stir things up before beginning to negotiate for your true interests.
14	Raise a Corpse from the Dead	Revive a dead proposal by presenting it again or in a new way.
15	Lure the Tiger out of the Mountain	Seek a neutral location. Negotiate after leading your enemy away from a position of strength.
16	To Catch Something, First Let It Go	Do not provoke your enemy's spirit to fight back.
17	Toss out a Brick to Attract a Piece of Jade	Trade something of minor value for something of major value.
18	To Catch Bandits, Nab Their Ringleader First	Convince the leader and the rest will follow.
19	Remove the Fire from under the Cauldron	Eliminate the source of your enemy's strength.
20	Muddle the Water to Catch the Fish	Do something surprising or unexpected to unnerve them, and then take advantage of that situation.
21	The Cicada Sheds Its Shells	When you are in trouble, escape in secret.
22	Fasten the Door to Catch a Thief	Annihilate your enemy by leaving no way for escape.
23	Befriend a Distant State While Attacking a Neighboring State	Build strategic alliances with others that will give you a strategic advantage.
24	Borrow a Safe Passage to Conquer the Kingdom of Guo	Temporarily join forces with another against a common enemy.
25	Steal the Beams and Pillars and Replace Them with Rotten Timber	Sabotage, incapacitate, or destroy the enemy by removing his key support.

	The Original 36 Stratagems	Contemporary Maxims
26	Point at the Mulberry Tree but Curse the Locust Tree	Convey your intentions and opinions indirectly.
27	Feign Ignorance and Hide One's Intentions	Play dumb, then surprise your enemy. Let him underestimate you.
28	Remove the Ladder after your Ascent	Lead the enemy into a trap, then cut off his escape.
29	Decorate the Tree with Fake Blossoms	Reframe deceitfully. Offer concessions with objects of misleading value.
30	Turn Yourself into a Host from Being a Guest	Turn your defensive and passive position into an offensive and active one.
31	Use a Beauty to Ensnare a Man	Provide alluring distractions.
32	Open the Gate of an Undefended City	Carefully displaying your weakness can conceal your vulnerability.
33	Use Adversary's Spies to Sow Discord in Your Adversary's Camp	Provide inaccurate information to mislead your enemy, especially through informal channels.
34	Inflict Pain on Oneself in order to Infiltrate Adversary's Camp and Win the Confidence of the Enemy	Appear to take some hits. Feign weakness while arming yourself.
35	Lead Your Adversary to Chain Together Their Warships	Devise a set of interlocking stratagems to defeat your enemy.
36	If All Else Fails, Run Away	Live to fight another day.

ACKNOWLEDGMENTS

This book has been a journey lasting years, and I am grateful to all the fellow travelers along the way. First, to my wife, Jessica, who is no stranger to war: a West Pointer, a military intelligence officer with three combat tours, and an ISIS chaser. She's peered into war's outer rim and has seen where it's heading.

To William Olson, my friend and a mentor (although he prefers "adviser"), who suffers from Cassandra's curse and was duly eaten by the Blob. Olson's Rules are maxims for the perplexed in Washington, DC—which is all of us—and I've sneaked a few into this book. My friend Tim Challans broadened my mind to Eastern strategic thought and is one of the most radical West Pointers you will ever meet. He's the only other soldier I know who listed "Taoism" under religion on his dog tags. I am also grateful to Mike Bell, who revealed to me the United States' own forgotten history of unconventional war, the dominant feature of the American way of war until World War II.

I have many friends to thank at the College of International Security Affairs (CISA) and at the National Defense University (NDU). I'm an unrepentant believer in professional military education. I've gained much wisdom from its faculty and students alike. CISA is a new model for a war college, and I think it's the best in the world at strategic education. However, friends at other institutions have pointed out that I'm a wee bit biased.

I am especially indebted to Rob Johnson, Peter Wilson, and the fellows at Oxford's Changing Character of War Centre. To say they influenced my thinking on the future of war understates the case. Someone needs to clone this program and export it.

Heaps of gratitude to the Air Power Policy Seminar. This symposium of erudite minds helped shape many thoughts over the years at our Connecticut Avenue headquarters. Thank you also to colleagues at Georgetown University's School of Foreign Service who provided smart feedback to sharpen the ideas of the book.

Thank you to David Highfill, Peter McGuigan, Bret Witter, and the mysterious copy editor who asked me all the difficult questions that experts overlook. Their guidance and edits greatly improved this book.

The views expressed in this book are entirely my own and do not reflect the position of the US Department of Defense or of any US government entity.

NOTES

STRATEGIC ATROPHY

1 "Losing is hateful to an American": Michael Keane, *George S. Patton: Blood, Guts, and Prayer* (Washington, DC: Regnery History, 2012), 104.

3 US polls on Afghanistan: Baxter Oliphant, "The Iraq War Continues to Divide the US Public, 15 Years after It Began," Pew Research Center, 19 March 2018, www.pewresearch.org/fact-tank/2018/03/19/iraq-war-continues-to-divide-u-s-public-15-years-after-it-began; Susan Page, "Poll: Grim Assessment of Wars in Iraq, Afghanistan," *USA Today*, 31 January 2014, https://eu.usatoday.com/story/news/politics/2014/01/30/usa-today-pew-research-poll-americans-question-results-in-iraq-afghanistan/5028097.

3 UK polls on Afghanistan: Theo Farrell, "Britain's War in Afghanistan: Was It Worth It?" *The Telegraph*, 6 June 2018, www.telegraph.co.uk/news/worldnews/asia/afghanistan/11377817/Britains-war-in-Afghanistan-was-it-worth-it.html.

3 John McCain, Iraq was a "mistake": Michael Hirsh, "John McCain's Last Fight," *Politico Magazine*, 18 May 2018, www.politico.com/magazine/story/2018/05/18/john-mccains-last-fight-218404.

3 “rules” versus “laws” of war: When I discuss the New Rules of War, I am not referring to the Law of Armed Conflict or various other “laws of war” that attempt to govern such behavior on the battlefield as killing civilians. This book brackets the important question of war ethics, and it takes as its starting point the decision to fight a war, not the question of whether one should do so.

5 French minds “were too inelastic”: Marc Bloch, *Strange Defeat* (Oxford: Oxford University Press, 1949), 36–37, 45.

5 “Strategic atrophy”: James N. Mattis, Secretary of Defense and retired four star general, also uses this term to voice concern. In the unclassified synopsis of the classified 2018 National Defense Strategy, he explains “we are emerging from a period of strategic atrophy, aware that our competitive military advantage has been eroding. We are facing increased global disorder, characterized by decline in the long-standing rules-based international order.” See: *Summary of the 2018 National Defense Strategy of the United States of America: Sharpening the American Military’s Competitive Edge.* Washington, DC: Department of Defense of the United States of America, 2018, 1, https://dod.defense.gov/Portals/1/Documents/pubs/2018-National-Defense-Strategy-Summary.pdf.

7 The world was safer during the Cold War: “COW War Data, 1817–2007 (v. 4.0),” The Correlates of War Project, 8 December 8, 2011, www.correlatesofwar.org/data-sets. See also Meredith Reid Sarkees and Frank Wayman, *Resort to War: 1816–2007* (Washington, DC: CQ Press, 2010); *The Fragile States Index*, Fund for Peace, 2018, http://fundforpeace.org/fsi; *Conflict Barometer 2017*, The Heidelberg Institute for International Conflict Research, 31 December, 2017, https://hiik.de/2018/02/28/conflict-barometer-2017/?lang=en; David Backer, Ravinder Bhavnani, and Paul Huth, eds., *Peace and Conflict 2016*

(Abingdon, UK: Routledge, 2016); Therése Pettersson and Peter Wallensteen, "Armed Conflicts, 1946–2014," *Journal of Peace Research* 52, no. 4 (2015): 536–50.

8 Wars no longer end but smolder in perpetuity: Matthew Hoddie and Caroline Hartzell, "Civil War Settlements and the Implementation of Military Power-Sharing Arrangements," *Journal of Peace Research* 40, no. 3 (2003): 303–20; Monica Duffy Toft, *Securing the Peace: The Durable Settlement of Civil Wars* (Princeton: Princeton University Press, 2009); Ben Connable and Martin C. Libicki, *How Insurgencies End*, vol. 965 (Santa Monica: RAND Corporation, 2010); Roger Mac Ginty, "No War, No Peace: Why So Many Peace Processes Fail to Deliver Peace," *International Politics* 47 no. 2 (2010): 145–62; Jasmine-Kim Westendorf, *Why Peace Processes Fail: Negotiating Insecurity after Civil War* (Boulder: Lynne Rienner Publishers, 2015).

WHY DO WE GET WAR WRONG?

11 War futurists are almost always wrong: Lawrence Freedman, *The Future of War: A History* (New York: Public Affairs, 2017).

14 CIA failed to see the collapsing USSR: Elaine Sciolino, "Director Admits CIA Fell Short in Predicting the Soviet Collapse," *New York Times*, 21 May 1992, www.nytimes.com/1992/05/21/world/director-admits-cia-fell-short-in-predicting-the-soviet-collapse.html.

14 U.S. Army "Iron Man" suit programs: Some fear $80 million is far too modest for TALOS. In the 1990s, the army spent $500 million on something similar called the Land Warrior, which didn't work. See Matt Cox, "Congress Wants More Control of Special Ops Iron Man Suit," Military.com, 29 April 2014, www.military.com/defensetech/2014/04/29/congress-wants-more-control-of-special-ops-iron-man-suit; Matthew

Cox, "Industry: Iron Man Still Hollywood, Not Reality," Military.com, 7 June 2018, www.military.com/daily-news/2014/04/22/industry-iron-man-still-hollywood-not-reality.html.

15 Rise of the robots: Matthew Rosenberg and John Markoff, "The Pentagon's 'Terminator Conundrum': Robots That Could Kill on Their Own," *New York Times*, 25 October 2016, www.nytimes.com/2016/10/26/us/pentagon-artificial-intelligence-terminator.html; Kevin Warwick, "Back to the Future," *Leviathan*, BBC News, 1 January 2000, http://news.bbc.co.uk/hi/english/static/special_report/1999/12/99/back_to_the_future/kevin_warwick.stm.

15 Robots are stupid: Andrej Karpathy and Li Fei-Fei, "Deep Visual-Semantic Alignments for Generating Image Descriptions," *Proceedings of the IEEE Conference on Computer Vision and Pattern Recognition* (2015): 3128–37, http://cs.stanford.edu/people/karpathy/cvpr2015.pdf.

15 What is "cyber"?: Cyber is a prefix used to describe anything having to do with computers, which doesn't explain much. The term "cyber" was coined by the science fiction writer William Gibson in the 1980s but has advanced little as a concept since then. Another example of life imitating art. William Gibson, *Burning Chrome* (New York: Ace Books, 1987).

15 CIA director warns of cyber "Pearl Harbor": Jason Ryan, "CIA Director Leon Panetta Warns of Possible Cyber–Pearl Harbor," *ABC News* 11 February 2011, http://abcnews.go.com/News/cia-director-leon-panetta-warns-cyber-pearl-harbor/story?id=12888905.

16 Inflated cyber threats to U.S. electrical grid: *Transforming the Nation's Electricity System: The Second Installment of the Quadrennial Energy Review* (Washington, DC: Department of Energy, January 2017), S-15. On varmint threat, see: Cyber Squirrel 1, 31 January 2018, http://cybersquirrel1.com.

16 Stuxnet hype: Michael Joseph Gross, "A Declaration of Cyber-War," *Vanity Fair*, 21 March 2011, www.vanityfair.com/news/2011/03/stuxnet-201104; Kim Zetter, "An Unprecedented Look at Stuxnet, the World's First Digital Weapon," *Wired*, 3 November 2014, www.wired.com/2014/11/countdown-to-zero-day-stuxnet.

17 Billy Mitchell predicts age of air power: William Mitchell, *Winged Defense: The Development and Possibilities of Modern Air Power—Economic and Military* (New York: G. P. Putnam's Sons, 1924), 25–26.

18 Billy Mitchell predicts Pearl Harbor: "Billy Mitchell's Prophecy," *American Heritage* 13, no. 2 (February 1962): www.americanheritage.com/content/billy-mitchell's-prophecy.

21 Fuller predicts age of *mechanized warfare:* John Frederick Charles Fuller, *On Future Warfare* (London: Sifton, Praed & Company, 1928).

21 Olson predicts the post-9/11 world: William J. Olson, "Global Revolution and the American Dilemma," *Strategic Review* (Spring 1983): 48–53.

22 Dereliction of duty: When generals have spoken out against the wars, it has been only after they've retired. In written memoirs or in media interviews, many have said that they doubted Rumsfeld's plan all along; meanwhile, they continued to send troops to their deaths. Given their later statements, such actions represent a moral failure to stand by their principles. Besides Shinseki, a notable exception to this grim trend was Marine lieutenant general Gregory Newbold, who openly criticized Rumsfeld's plans for Iraq and retired partly in protest in 2002.

22 Science fiction presented as "research": The Art of the Future Project, The Atlantic Council, Brent Scowcroft Center on International Security, http://artoffuturewarfare.org.

24 A $12.9 billion ship: Kris Osborn, "In Quest for 355 Ships, Navy May Buy Two Supercarriers At Once," Military.com, 20 March 2018, www

.military.com/dodbuzz/2018/03/20/quest-355-ships-navy-may-buy-two-supercarriers-once.html.

RULE 1: CONVENTIONAL WAR IS DEAD

25 Civilian massacre: William Caferro, *John Hawkwood: An English Mercenary in Fourteenth-Century Italy* (Baltimore: JHU Press, 2006), 189.

27 1300 CE, or today: BCE stands for "Before Common Era" and CE means "Common Era." They are alternatives to BC ("Before Christ") and AD ("Anno Domini"). The two notation systems are equivalent, so 400 BCE is the same as 400 BC. Scholars prefer BCE and CE because they are religiously neutral.

28 "Conventional war": Technically, it should be "conventional warfare," not "conventional war." However, usage trumps grammar and everyone calls it "conventional war," so I will stick with this convention.

28 World War II movies: Internet Movie Database (IMDb), https://www.imdb.com.

29 On Clausewitz: Despite this reverence, Clausewitz is the most quoted yet least understood man in the Pentagon. Another seminal thinker of conventional war is Antoine-Henri Jomini (1779–1869), a Swiss Napoleonic general who published *The Art of War* in 1838. Jomini once eclipsed Clausewitz in influence but has since fallen out of favor. Jomini's and Clausewitz's ideas have more in common than not, and they have shaped much of Western strategic thought. Clausewitz gives us a strategic lexicon for war, but his relevance remains hotly debated. For example, pro: Christopher Coker, *Rebooting Clausewitz: 'On War' in the Twenty-First Century* (New York: Oxford University Press, 2017); con: William J. Olson, "The Continuing Irrelevance of Clausewitz,"

Small Wars Journal, 26 July 2013, http://smallwarsjournal.com/jrnl/art/the-continuing-irrelevance-of-clausewitz.

30 Terms used interchangeably: This book treats "conventional," "regular," and "symmetrical" war as the same. Similarly, it uses the terms "nation-state" and "state" interchangeably.

31 Peace of Westphalia: The consensus view among political scientists is that the Peace of Westphalia represents the birth of the modern world order. However, the language of the treaties of Münster and Osnabrück, which compose the Peace, do not articulate an international system of states. Rather, they are merely a ceasefire that twentieth-century scholars have reified into fact. For more, see Stephen D. Krasner, "Westphalia and All That," in *Ideas and Foreign Policy: Beliefs, Institutions, and Political Change*, ed. Judith Goldstein and Robert O. Keohane (Cornell, NY: Cornell University Press, 1993), 235; Stephen D. Krasner, *Sovereignty: Organized Hypocrisy* (Princeton, NJ: Princeton University Press, 1999), 82; Edward Keene, *Beyond the Anarchical Society: Grotius, Colonialism and Order in World Politics* (Cambridge: Cambridge University Press, 2002); Andreas Osiander, "Sovereignty, International Relations, and the Westphalian Myth," *International Organization* 55, no. 2 (2001): 284; Benno Teschke, *The Myth of 1648: Class, Geopolitics, and the Making of Modern International Relations* (New York: Verso, 2003).

32 Fragile States Index: J. J. Messner et al., *Fragile States Index 2017–Annual Report* (Washington, DC: Fund for Peace, 14 May 2017), http://fundforpeace.org/fsi/2017/05/14/fragile-states-index-2017-annual-report.

35 Chart data: Uppsala Conflict Data Program (UCDP)/Peace Research Institute Oslo (PRIO) Armed Conflict Dataset, version 17.2. (Note: This data set includes only armed conflicts that involved at least one

state and disregards conflicts between nonstate actors alone.) See also Marie Allansson, Erik Melander, and Lotta Themnér, "Organized Violence, 1989–2016," *Journal of Peace Research* 54, no. 4 (2017); Nils Petter Gleditsch et al., "Armed Conflict, 1946–2001: A New Dataset," *Journal of Peace Research* 39, no. 5 (2002): 615–37.

35 Armed conflict trends: Kendra Dupuy et al., *Trends in Armed Conflict, 1046–2015* (Oslo: Peace Research Institute Oslo, August 2016), www.prio.org/utility/DownloadFile.ashx?id=15&type=publicationfile. For other studies, see "COW War Data, 1817–2007 (v 4.0)," The Correlates of War Project, 8 December 2011, www.correlatesofwar.org/data-sets; Meredith Reid Sarkees and Frank Wayman, *Resort to War: 1816–2007* (Washington DC: CQ Press, 2010); Uppsala Conflict Data Program (UCDP)/Peace Research Institute Oslo (PRIO) Armed Conflict Dataset, 31 December 2016, www.ucdp.uu.se/gpdatabase/search.php. See also Gleditsch et al., "Armed Conflict 1946–2001"; Kalevi Holsti, *Kalevi Holsti: Major Texts on War, the State, Peace, and International Order* (Springer International Publishing, 2016).

36 "The days of armed conflict between nation-states are ending": See Tom Ricks, "U.S. Faces Defense Choices: Terminator, Peacekeeping Globocop or Combination," *Wall Street Journal*, 12 November 1999, www.wsj.com/articles/SB942395118449418125.

37 US buying more submarines: US Congress, Senate, House, *National Defense Authorization Act For Fiscal Year 2018*, HR 2810, 115th Cong., 1st sess., enacted 3 January 2017, www.congress.gov/115/bills/hr2810/BILLS-115hr2810enr.pdf.

38 Annual budget of US special operations forces: US Congress, House, Committee on Armed Services, Subcommittee on Emerging Threats and Capabilities, *The Future of U.S. Special Operations Forces: Hearing before the Subcommittee on Emerging Threats and Capabilities*, 112th

Cong., 2nd sess., 11 July 2012, www.gpo.gov/fdsys/pkg/CHRG-112 hhrg75150/pdf/CHRG-112hhrg75150.pdf. See also Linda Robinson, *The Future of US Special Operations Forces*, Council Special Report no. 66 (New York: Council on Foreign Relations, 2013).

39 US military reserves cannot keep up with war requirements: Office of the Secretary of Defense, *Final Report of the Defense Science Board Task Force on Deployment of Members of the National Guard and Reserve in the Global War on Terrorism*, September 2007, www.acq.osd.mil/dsb /reports/2000s/ADA478163.pdf.

42 James Mattis Testimony: US Congress, Senate, Armed Services Committee, *Defense Authorization for Central Command and Special Operations*, 3 March 2013, testimony by James N. Mattis, commander (former) US Central Command. Mattis's testimony was recorded by C-SPAN; a user-created clip posted online on 28 February 2017 is available at www.c-span.org/video/?c4658822/mattis-ammunition.

RULE 2: TECHNOLOGY WILL NOT SAVE US

44 F-35 a "needed aircraft": Mark Thompson, "The Most Expensive Weapon Ever Built," *Time*, 25 February 2013, http://content.time .com/time/printout/0,8816,2136312,00.html; Amanda Macias, "We Spent a Day with the People Who Fly and Fix the F-35–Here's What They Have to Say about the Most Expensive Weapons Project in History," *Business Insider*, 26 September 2016, www.businessinsider.com /f35-pilot-interview-2016-9.

44 F-35 costs: Thompson, "The Most Expensive Weapon"; James Drew, "F-35A Cost and Readiness Data Improves in 2015 as Fleet Grows," FlightGlobal.com, 2 February 2016, www.flightglobal.com/news /articles/f-35a-cost-and-readiness-data-improves-in-2015-as-fl-421499.

45 A-10 outperforms F-35: David Cenciotti, "'A-10 Will Always Be Better Than F-35 in Close Air Support. In All the Other Missions the JSF Wins,' F-35 Pilot Says," *The Aviationist*, 9 April 2015, https://theaviationist.com/2015/04/09/f-35-never-as-a10-in-cas.

45 F-15 and F-16 outperform F-35: David Axe, "Test Pilot Admits the F-35 Can't Dogfight," *War Is Boring*, 29 June 2015, https://warisboring.com/test-pilot-admits-the-f-35-can-t-dogfight.

45 F-35 buggy computer code: Thompson, "The Most Expensive Weapon"; Clay Dillow, "Pentagon Report: The F-35 Is Still a Mess," *Fortune*, 10 March 2016, http://fortune.com/2016/03/10/the-f-35-is-still-a-mess.

45 "Category I" errors: "F-35 Joint Strike Fighter (JSF)," FY16 Department of Defense Programs, 2015, p. 49, www.dote.osd.mil/pub/reports/FY2015/pdf/dod/2015f35jsf.pdf.

46 "acquisitions malpractice": Thompson, "The Most Expensive Weapon."

46 James Mattis Testimony: US Congress, Senate, Committee on Armed Services, *Global Challenges, US National Security Strategy and Defense Organization: Hearings before the Committee on Armed Services*, 114th Cong., 1st sess., 21, 27, 29 January and 22 October 2015, https://archive.org/details/gov.gpo.fdsys.CHRG-114shrg22944.

48 Third Offset Strategy: Robert O. Work, "The Third US Offset Strategy and Its Implications for Partners and Allies" (remarks delivered at the Willard Hotel, Washington, DC, 28 January 2015), www.defense.gov/News/Speeches/Speech-View/Article/606641/the-third-us-offset-strategy-and-its-implications-for-partners-and-allies.

49 "no greater sin in the profession": Sydney J. Freedberg Jr., "War Without Fear: DepSecDef Work on How AI Changes Conflict," *Breaking Defense*, 31 May 2017, https://breakingdefense.com/2017/05/killer-robots-arent-the-problem-its-unpredictable-ai.

50 "friggin' robot": Cheryl Pellerin, "Work: Human-Machine Teaming

Represents Defense Technology Future," *Department of Defense News*, 8 October 2015, www.defense.gov/News/Article/Article/628154/work-human-machine-teaming-representsdefense-technology-future.

50 Defense Innovation Unit-Experimental or DIUx: Franz-Stefan Gady, "New US Defense Budget: $18 Billion for Third Offset Strategy," *The Diplomat*, 10 February 2016, http://thediplomat.com/2016/02/new-us-defense-budget-18-billion-for-third-offset-strategy.

50 Bob Work joins Raytheon: "Robert O. Work Elected to Raytheon Board of Directors," CISION PR Newswire, 14 August 2017, www.prnewswire.com/news-releases/robert-o-work-elected-to-raytheon-board-of-directors-300503294.html; Aaron Mehta, "The Top 100: A Return to Prosperity?," *Defense News*, 20 July 2017, www.defensenews.com/2017/07/20/finally-defense-revenues-grow-for-first-time-in-five-years.

50 Project Maven: Office of the Deputy Secretary of Defense, "Establishment of an Algorithmic Warfare Cross-Functional Team (Project Maven)," memorandum, 26 April 2017, www.govexec.com/media/gbc/docs/pdfs_edit/establishment_of_the_awcft_project_maven.pdf.

51 Google resignations over Project Maven: Scott Shane et al., "'The Business of War': Google Employees Protest Work for the Pentagon," *New York Times*, 4 April 2018, www.nytimes.com/2018/04/04/technology/google-letter-ceo-pentagon-project.html.

53 USS *Fitzgerald* collision: Tom Vanden Brook, "Commander and Leadership of Stricken Destroyer Fitzgerald to Be Relieved for Collision," *USA Today*, 19 August 2017.

55 "eyes on the iron": Mark D. Faram, "Maybe Today's Navy Is Just Not Very Good at Driving Ships," *Navy Times*, 27 August 2017, www.navytimes.com/news/your-navy/2017/08/27/navy-swos-a-culture-in-crisis.

56 100+ hour workweek aboard navy ships: US Congress, House, Com-

mittee on Armed Services, Subcommittees on Readiness and Seapower and Projection Forces, *Navy Readiness: Actions Needed to Address Persistent Maintenance, Training, and Other Challenges Facing the Fleet*, 115th Cong., 1st sess., 7 September 2017, testimony of John H. Pendleton, director, Defense Capabilities and Management, released online by the Government Accountability Office, GAO-17-798T, 7 September 2017, www.gao.gov/assets/690/686995.pdf.

56 US naval officer gets schooled by British navy: Mitch McGuffie, "A Rude Awakening," *Proceedings* (US Naval Institute) 135, no. 1 (January 2009), https://www.usni.org/magazines/proceedings/2009-01/rude-awakening.

RULE 3: THERE IS NO SUCH THING AS WAR OR PEACE—BOTH COEXIST, ALWAYS

63 USS *Stethem*: This passage is a speculative reconstruction and not an eyewitness account of actions aboard the USS *Stethem* on July 2, 2017. It is based on an interview with a US navy commanding officer of an Arleigh Burke destroyer who conducted Freedom of Navigation Operations (FONOPs) missions in the South China Sea in 2017, just as the *Stethem* did.

64 Russian "New Generation Warfare": A term devised by Western analysts, who also label Russia's new way of war the "Gerasimov Doctrine" or "Hybrid Warfare." Russians rarely use these terms. For more, see Dmitry Adamsky, "From Moscow with Coercion: Russian Deterrence Theory and Strategic Culture," *Journal of Strategic Studies* 41, no. 1–2 (2018): 39.

64 Israel's new war strategy: "Deterring Terror: How Israel Confronts the Next Generation of Threats," English translation of the *Official Strat-*

egy of the Israel Defense Forces, Harvard Kennedy School Belfer Center for Science and International Affairs, special report, August 2016, www.belfercenter.org/sites/default/files/legacy/files/IDFDoctrineTranslation.pdf.

65 "a few scattered rocks in the Pacific": Craig Whitlock, "Panetta to Urge China and Japan to Tone Down Dispute," *Washington Post*, 16 September 2012, www.washingtonpost.com/world/national-security/panetta-to-urge-china-and-japan-to-tone-down-dispute-over-islands/2012/09/16/9b6832c0-fff3-11e1-b916-7b5c8ce012c8_story.html.

66 China's "Three Warfares" strategy: Stefan Halper, *China: The Three Warfares*, report for the Office of Net Assessment, US Department of Defense, 2013, p. 11, available online at https://cryptome.org/2014/06/prc-three-wars.pdf.

67 "use every arrow": Michael Martina, "US Business Group Urges Washington to 'Use Every Arrow' Against China," Reuters, 18 April 2017, www.reuters.com/article/us-china-usa-business/u-s-business-group-urges-washington-to-use-every-arrow-against-china-idUSKBN17K0G8.

67 Chamber of Commerce spent $82 million on lobbying: "Annual Lobbying by the U.S. Chamber of Commerce 2017," Center for Responsible Politics, accessed 8 June 2018, www.opensecrets.org/lobby/clientsum.php?id=D000019798&year=2017.

67 China's state controlled media: "China's CCTV Launches Global 'Soft Power' Media Network to Extend Influence," Reuters, 31 December 2016, published online by *Fortune*, http://fortune.com/2016/12/31/chinas-cctv-global-influence.

67 China buys Hollywood: Anita Busch and Nancy Tartaglione, "China & Hollywood: What Lies Beneath & Ahead in 2017," *Deadline*, 5 Jan-

uary 2017, http://deadline.com/2017/01/china-hollywood-deals-2017-donald-trump-1201875991.

68 CCTV spin blames right-wing nationalists: "China's Non-Kinetic 'Three Warfares' Against America," *The National Interest*, 5 January 2016, http://nationalinterest.org/print/blog/the-buzz/chinas-non-kinetic-three-warfares-against-america-14808.

68 The "Big Lie": Adolf Hitler, *Mein Kampf*, trans. James Murphy, available online at Project Gutenberg Australia, last updated September 2002, http://gutenberg.net.au/ebooks02/0200601.txt.

68 Chinese "lawfare": Major General Liu Jiaxin, "General's Views: Legal Warfare—Modern Warfare's Second Battlefield," *Guangming Ribao*, 3 November 2004. At the time, Liu was commandant of the Xian Political Academy of the PLA General Political Department.

69 China rules the cosmos too: Orde F. Kittrie, *Lawfare: Law as a Weapon of War* (New York: Oxford University Press, 2016), 168.

69 Future war fantasy: Peter W. Singer and August Cole, *Ghost Fleet: A Novel of the Next World War* (New York: Houghton Mifflin Harcourt, 2015). See also http://www.marines.mil/News/Messages/Messages-Display/Article/1184470/revision-of-the-commandants-professional-reading-list.

70 "acts of war" that do not lead to war: Joshua Keating, "Why Are Nations Rushing to Call Everything an 'Act of War'?," *New York Times Magazine*, 12 December 2017, www.nytimes.com/2017/12/12/magazine/why-are-nations-rushing-to-call-everything-an-act-of-war.html.

76 "One prominent scholar": Hal Brands. *What Good Is Grand Strategy?: Power and Purpose in American Statecraft From Harry S. Truman to George W. Bush* (New York, NY: Cornell University Press, 2014).

RULE 4: HEARTS AND MINDS DO NOT MATTER

83 Jewish messiahs: Other "messiahs" of this time included Judas, son of Ezekias; Simon, a slave of King Herod; Anthronges, a shepherd; Menahem, a descendent of Judas of Galilee; John of Gischala; and Simon bar Giona. Additionally, several popular messianic rebellions sprang up spontaneously at the death of Herod in 4 BCE.

87 Roman siege of Jerusalem: Flavius Josephus, *The Jewish War*, trans. G. A. Williamson (New York: Penguin Books, 1970), 312-324; Max I. Dimont, *Jews, God, and History* (New York: Penguin Books), 101.

87 Massacre at Ein Gedi: Flavius Josephus, *The Jewish War*, trans. G. A. Williamson (New York: Penguin Books, 1970), 255.

89 Eleazar's last speech: Flavius Josephus, *The Jewish War*, trans. G. A. Williamson (New York: Penguin Books, 1970), 385-90.

90 "War is hell": William T. Sherman, *"War Is Hell!": William T. Sherman's Personal Narrative of His March Through Georgia* (Savannah: Beehive Press, 1974).

92 T. E. Lawrence explains guerrilla warfare: Thomas Edward Lawrence, *Evolution of a Revolt*, originally published in *Army Quarterly* 1, no. 1 (October 1920), published in the United States by Praetorian Press in 2011.

93 Hearts and minds: Letter from John Adams to Hezekiah Niles on the American Revolution, 13 February 1818, http://nationalhumanities center.org/ows/seminars/revolution/Adams-Niles.pdf.

93 "Military action is secondary to the political one": David Galula, *Counterinsurgency Warfare: Theory and Practice* (Westport, CT: Praeger Security International, 1964), 63.

94 "armed social work": David Kilcullen, "Twenty-Eight Articles: Funda-

mentals of Company-Level Counterinsurgency," *Military Review* 86, no. 3 (May–June 2006): 105–7.

96 Clausewitz on counterinsurgency: Clausewitz, like nearly all military thinkers of his age, did not consider insurgency war. Counterinsurgency meant putting the rebels to the sword. This was generally not a problem since, in Clausewitz's opinion, rebellion was a capital offense and peasants were no match for professional soldiers. Most of history would agree. See *On War*, 479–83.

96 "as a fish swims in the sea": Mao Tse-tung [Mao Zedong], *On Guerrilla Warfare*, trans. Samuel B. Griffith II (Chicago: University of Illinois Press, 2000).

97 China's "peaceful liberation" of Tibet: "Peaceful Liberation of Tibet," *Xinhua*, 22 May 2001, accessed at China.org on 8 June 2018, www.china.org.cn/english/13235.htm.

99 American vets struggle to join Foreign Legion: Slobodan Lekic, "Americans Struggle to Meet the French Foreign Legion's High Bar," *Stars and Stripes*, 12 November 2007, www.stripes.com/news/europe/americans-struggle-to-meet-the-french-foreign-legion-s-high-bar-1.497591.

100 US citizenship through military service: US Citizenship and Immigration Services, *Naturalization Through Military Service: Fact Sheet*, accessed 8 June 2018, www.uscis.gov/news/fact-sheets/naturalization-through-military-service-fact-sheet.

101 CIA and Pentagon militias battle each other: W. J. Hennigan et al., "In Syria, Militias Armed by the Pentagon Fight Those Armed by the CIA," *Los Angeles Times*, 27 March 2016, www.latimes.com/world/middleeast/la-fg-cia-pentagon-isis-20160327-story.html.

101 US trained militias gave their weapons to ISIS: Nabih Bulos, "US-Trained Division 30 Rebels 'Betray US and Hand Weapons Over to

al-Qaeda's Affiliate in Syria," *The Telegraph*, 22 September 2015, www.telegraph.co.uk/news/worldnews/middleeast/syria/11882195/US-trained-Division-30-rebels-betrayed-US-and-hand-weapons-over-to-al-Qaedas-affiliate-in-Syria.html.

101 US spent $500 million to create a 5-person militia: "Syria Crisis: 'Only Four or Five' US-Trained Syrian Rebels Are Still Fighting," BBC News, 17 September 2015, www.bbc.com/news/world-middle-east-34278233.

101 "heaps of problems" created by contractors: Problems created by contractors will be explained in "Rule 6: Mercenaries Will Return." On contractor numbers during the Iraq and Afghanistan wars, see: Heider M. Peters et al., *Department of Defense Contractor and Troop Levels in Iraq and Afghanistan: 2007–2017*, US Library of Congress, Congressional Research Service, R44116, 28 April 2017, p. 4, https://fas.org/sgp/crs/natsec/R44116.pdf; Office of the Assistant Secretary of Defense for Logistics and Materiel Readiness, *Past Contractor Support of U.S. Operations in USCENTCOM AOR, Iraq, and Afghanistan: Quarterly Contractor Census Reports, 2008–2018*, April 2018, p. 4, www.acq.osd.mil/log/PS/CENTCOM_reports.html.

102 Fraud, waste, and abuse in Iraq and Afghanistan: Special Inspector General for Iraq Reconstruction, *Learning from Iraq: A Final Report*, March 2013, www.globalsecurity.org/military/library/report/2013/sigir-learning-from-iraq.pdf; US Congress, Senate, Committee on Armed Services, Subcommittee on Readiness and Management Support, *The Final Report of the Commission on Wartime Contracting in Iraq and Afghanistan: Hearing before the Subcommittee on Readiness and Management Support*, 112th Cong., 1st sess., 19 October 2011, www.gpo.gov/fdsys/pkg/CHRG-112shrg72564/pdf/CHRG-112shrg72564.pdf.

RULE 5: THE BEST WEAPONS DO NOT FIRE BULLETS

104 Zapad-81: Central Intelligence Agency, *Planning, Preparation, Operation and Evaluation of Warsaw Pact Exercises*, 1981, approved for release by the Historical Collection Division, 18 June 2012, https://www.cia.gov/library/readingroom/docs/1981-01-01.pdf.

105 Germany spent $6.7 billion resettling refugees: Thomas de Maiziere, "965.000 Flüchtlinge bis Ende November in Deutschland," *Welt*, 7 December 2015, www.welt.de/politik/deutschland/article149700433/965-000-Fluechtlinge-bis-Ende-November-in-Deutschland.html; Holly Yan, "Are Countries Obligated to Take In Refugees? In Some Cases, Yes," CNN, 29 December 2015, http://edition.cnn.com/2015/09/08/world/refugee-obligation/index.html.

106 Philip Breedlove testimony: US Congress, Senate, Committee on Armed Services, *Hearing to receive testimony on United States European Command*, Committee on Armed Services, 114th Cong., 2nd sess., 1 March 2016, www.armed-services.senate.gov/imo/media/doc/16-20_03-01-16.pdf.

107 "future threats are not conquering states but failing ones": Even the United States' national-security establishments admit this to be true, although they have not metabolized its significance. See George W. Bush, "The National Security Strategy of the United States of America," The White House, United States Government, 2002, p. 1.

110 "We're being out-communicated by a guy in a cave": Karen DeYoung et al., "U.S. to Fund Pro-American Publicity in Iraqi Media," *Washington Post*, 3 October 2008, www.washingtonpost.com/wp-dyn/content/article/2008/10/02/AR2008100204223.html; Mark Mazzetti and Michael R. Gordon, "ISIS Is Winning the Social Media War, US Concludes,"

New York Times, 12 June 2015, www.nytimes.com/2015/06/13/world/middleeast/isis-is-winning-message-war-us-concludes.html.

RULE 6: MERCENARIES WILL RETURN

122 Haditha massacre: "'Simple Failures' and 'Disastrous Results'—Excerpts from Army Maj. Gen. Eldon A. Bargewell's Report: The Response to the Haditha incident," *Washington Post*, 21 April 2007, www.washingtonpost.com/wp-dyn/content/article/2007/04/20/AR2007042002309.html.

123 Machiavelli on mercenaries: Niccolò Machiavelli, *The Prince and Other Works* (New York: Hendricks House, 1964), 131.

124 Critiques of Machiavelli: Quentin Skinner, *Machiavelli: A Very Short Introduction* (New York: Oxford Paperbacks, 2000), 36–37; Christopher Coker, *Barbarous Philosophers: Reflections on the Nature of War From Heraclitus to Heisenberg* (New York: Columbia University Press, 2010), 139–51; James Jay Carafano, *Private Sector, Public Wars: Contractors in Combat—Afghanistan, Iraq, and Future Conflicts* (Westport, CT: Praeger Security International, 2008), 19; Sarah Percy, *Mercenaries: The History of a Norm in International Relations* (New York: Oxford University Press, 2007).

126 Biblical mercenaries: Herbert G. May, Bruce M. Metzger, eds., *The New Oxford Annotated Bible with the Apocrypha: Revised Standard Version, Containing the Second Edition of the New Testament and an Expanded Edition of the Apocrypha* (New York: Oxford University Press, 1977), Jeremiah 46:20–21; 2 Samuel 10:6; 1 Chronicles 19:7; 2 Kings 11:4; 2 Chronicles 25:6; 2 Kings 7:6–7; 2 Samuel 10:6; 1 Chronicles 19:6–7; 2 Chronicles 25:5–6;). See also *The Ryrie Study Bible*, New International Version (Chicago: Moody Publishers, 1994).

127 "Kill them all, God will know his own": Caesarius of Heisterbach, *Dialogue on Miracles V*, Chapter XXI, "Of the Heresy of the Albigenses" (c. 1223), edited by Paul Halsall, Fordham University Center for Medieval Studies. https://sourcebooks.fordham.edu/source/caesarius-heresies.asp.

128 Troop to contractor ratios: Heider M. Peters et al., *Department of Defense Contractor and Troop Levels in Iraq and Afghanistan: 2007–2017*, R44116 (Washington: US Library of Congress, Congressional Research Service, 28 April 2017), 4, https://fas.org/sgp/crs/natsec/R44116.pdf; *Past Contractor Support of U.S. Operations in USCENTCOM AOR, Iraq, and Afghanistan: Quarterly Contractor Census Reports 2008–2018* (Washington: Office of the Assistant Secretary of Defense for Logistics and Materiel Readiness, April 2018), 4, https://www.acq.osd.mil/log/PS/CENTCOM_reports.html.

129 Percentage of armed contractors: Peters et al., *Department of Defense Contractor and Troop Levels.*

129 Erik Prince's plan to privatize the Afghanistan war: Erik D. Prince, "The MacArthur Model for Afghanistan," *Wall Street Journal*, 31 May 2017, www.wsj.com/articles/the-macarthur-model-for-afghanistan-1496269058; On Prince's plan to foot the bill, see Aram Roston, "Private War: Erik Prince Has His Eye on Afghanistan's Rare Metals," *BuzzFeed News,* 7 December 2017, www.buzzfeed.com/aramroston/private-war-erik-prince-has-his-eye-on-afghanistans-rare.

Iraq war would "pay for itself": On the many experts who promised a cost-free war in Iraq, see Victor Navasky and Christopher Cerf, "Who Said the War Would Pay for Itself? They Did!" *The Nation*, 13 March 2008, www.thenation.com/article/who-said-war-would-pay-itself-they-did. On the actual cost of the Iraq War, see Neta C. Crawford, "US Budgetary Costs of Wars through 2016: $4.79 Trillion and Counting:

Summary of Costs of the US Wars in Iraq, Syria, Afghanistan and Pakistan and Homeland Security," Costs of War, September 2016, http://watson.brown.edu/costsofwar/files/cow/imce/papers/2016/Costs%20of%20War%20through%202016%20FINAL%20final%20v2.pdf.

129 Rumsfeld promised a quick war in Iraq: John Esterbrook, "Rumsfeld: It Would Be A Short War," CBS News, 15 November 2002, www.cbsnews.com/news/rumsfeld-it-would-be-a-short-war.

130 Contractor versus troop casualties: Steven L. Schooner and Collin Swan, "Contractors and the Ultimate Sacrifice," The George Washington University Law School, working paper no. 512 (September 2010); Christian Miller, "Civilian Contractor Toll in Iraq and Afghanistan Ignored By Defense Dept," ProPublica, 9 October 2009.

130 Contractor PTSD and casualties: Molly Dunigan et al., *Out of the Shadows: The Health and Well-Being of Private Contractors Working in Conflict Environments* (Santa Monica, CA: RAND Corporation, 2013); *Contingency Contracting: DOD, State, and U.S.AID Continue to Face Challenges in Tracking Contractor Personnel and Contracts in Iraq and Afghanistan*, GAO-10-1 (Washington: US Government Accountability Office, 2009), www.gao.gov/new.items/d101.pdf; Justin Elliott, "Hundreds of Afghanistan Contractor Deaths Go Unreported," *Salon*, 15 July 2010. "Statistics on the Private Security Industry," *Private Security Monitor*, University of Denver, accessed 13 June 2018, http://psm.du.edu/articles_reports_statistics/data_and_statistics.html.

131 Contractors cheaper than troops: US Congress, Congressional Budget Office, *Contractors' Support of U.S. Operations in Iraq*, 110th Cong., 2nd sess., August 2008, p. 17.

131 Contractors better than UN in Sierra Leone: Herbert Howe, "Private Security Forces and African Stability: The Case of Executive Outcomes," *Journal of Modern African Studies* 36, no. 2 (June 1998);

S. Mallaby, "New Role for Mercenaries," *Los Angeles Times*, 3 August, 2001; S. Mallaby, "Paid to Make Peace, Mercenaries Are No Altruists, but They Can Do Good," *Washington Post*, 4 June 2001.

131 US spent $160 billion on contractors: Moshe Schwartz and Jennifer Church, *Department of Defense's Use of Contractors to Support Military Operations: Background, Analysis, and Issues for Congress* (Washington: US Library of Congress, Congressional Research Service, 2013), 2; For the 2012 UK defense budget, see "UK Government Spending in 1998," UK Public Spending, accessed 6 June 2018, available online at www.ukpublicspending.co.uk/budget_current.php?title=uk_defense_budget&year=2012&fy=2012&expand=30.

131 Senator Obama on contractors: US Congress, Senate, *Transparency and Accountability in Military and Security Contracting Act of 2007*, bill introduced by Senator Barak Obama, 11th Cong., 1st sess., 16 February 2007, www.gpo.gov/fdsys/pkg/BILLS-110s674is/pdf/BILLS-110s674is.pdf.

132 Mercenaries battle elite US forces in Syria: Thomas Gibbons-Neff, "How a 4-Hour Battle Between Russian Mercenaries and US Commandoes Unfolded in Syria," *New York Times*, 24 May 2018, www.nytimes.com/2018/05/24/world/middleeast/american-commandos-russian-mercenaries-syria.html; "U.S. Forces Deploy to Conoco Gas Plant in Anticipation of Iranian Advance," *Alsouria Net*, 18 February 2018, reprinted at *The Syrian Observer* on 16 February 2018, http://syrianobserver.com/EN/News/33850/U_S_Forces_Deploy_Conoco_Gas_Plant_Anticipation_Iranian_Advance.

136 UN could outsource peacekeeping: To accomplish this requirement, the United Nations would have to establish a licensing and registration regime that all industry members would need to observe in order to prequalify for contracts with the organization. This would entail clear

standards and policies regulating all industry activities, plus mechanisms for oversight and accountability. As a minimum, this regime should include the following elements: registration criteria, ethical code of conduct, employee vetting standards, mechanisms of transparency and accountability, permissible clients (e.g., clients sanctioned by the UN Security Council), training and safety standards, contractual standards, and compliance enforcement mechanisms such as audits. Contract instruments would need to be in place to ensure the swift deployment of private military companies, should an emergency arise. It would be impermissible to lose a key advantage of the private sector's rapid response and surge capacity to bureaucratic dithering.

139 International law is the "vanishing point of law": Thomas E. Holland, *Elements of Jurisprudence,* 9th ed. (Oxford: Oxford University Press, 1900), 369.

RULE 7: NEW TYPES OF WORLD POWERS WILL RULE

143 World's largest megachurches: "Global Megachurches—World's Largest Churches," compiled and maintained by Warren Bird, Leadership Network, accessed 7 June 2018, http://leadnet.org/world.

144 US megachurches receive $6.5 million annually: "Mega Churches Mean Big Business," CNN, 22 January 2018, http://edition.cnn.com/2010/WORLD/americas/01/21/religion.mega.church.christian/index.html.

144 Lakewood Church's $90 million annual budget: Anugrah Kumar, "Joel Osteen's Lakewood Church Has Annual Budget of $90 Million: Here's How That Money Is Spent," *Christian Post,* 3 June 2018, www.christianpost.com/news/joel-osteen-lakewood-church-annual-budget-90-million-money-spent-224604.

145 Interview with modern crusader: Henry Tuck et al., "Shooting in the

Right Direction: Anti-ISIS Foreign Fighters in Syria and Iraq," Horizons Series, no. 1, Institute of Strategic Dialogue, August 2016, www.isdglobal.org/wp-content/uploads/2016/08/ISD-Report-Shooting-in-the-right-direction-Anti-ISIS-Fighters.pdf; Terry Moran, "Why This American 'Soldier of Christ' Is Fighting ISIS in Iraq," ABC News, 24 February 2015, https://abcnews.go.com/International/american-soldier-christ-fighting-isis-iraq/story?id=29171878; "Meet the 'Foreign Legion' of the Anti-ISIS Christian Militia," *Arutz Sheva*, 18 February 2015, www.israelnationalnews.com/News/News.aspx/191492.

146 Mia Farrow almost hired Blackwater: Harvey Morris, "Activists Turn to Blackwater for Darfur Help," *Financial Times*, 18 June 2008, https://www.ft.com/content/4699eda6-3d65-11dd-bbb5-0000779fd2ac.

149 "Dozens of studies confirm this": J. J. Messner et al., "Fragile States Index 2017—Annual Report," Fund for Peace, 14 May 2017, http://fundforpeace.org/fsi/2017/05/14/fragile-states-index-2017-annual-report. For examples of other fragile state rankings, see Daniel Kaufmann and Aart Kraay, "Worldwide Governance Indicators Project," World Bank, accessed 13 June 2018, http://info.worldbank.org/governance/wgi/#home; Arch Puddington and Tyler Roylance, "Populists and Autocrats: The Dual Threat to Global Democracy," *Freedom in the World Report* 17, Freedom House, https://freedomhouse.org/report/freedom-world/freedom-world-2017; *Human Development Report 2016*, United Nations Development Programme, http://hdr.undp.org/en/composite/HDI; "States of Fragility Reports," The Organization of Economic Co-operation and Development (OECD), last updated 2016, www.oecd.org/dac/conflict-fragility-resilience/listofstateoffragilityreports.htm; "Corruption Perception Index," Transparency International, 21 February 2018, www.transparency.org/research/cpi/overview. For examples of scholarship, see Robert I. Rotberg, "Failed

States, Collapsed States, Weak States: Causes and Indicators," in *State Failure and State Weakness in a Time of Terror* (Cambridge, MA: Brookings Institution Press, 2003), 1–25; Ashraf Ghani and Clare Lockhart, *Fixing Failed States: a Framework for Rebuilding a Fractured World* (New York: Oxford University Press, 2009); Jean-Germain Gros, "Towards a Taxonomy of Failed States in the New World Order: Decaying Somalia, Liberia, Rwanda and Haiti," *Third World Quarterly* 17, no. 3 (1996): 455–72.

150 Top and bottom ranked states: J. J. Messner et al., *Fragile States Index 2017–Annual Report* (Washington, DC: Fund for Peace, 14 May 2017), http://fundforpeace.org/fsi/2017/05/14/fragile-states-index-2017-annual-report.

151 Global 1% owns most everything: Anthony Shorrocks, Jim Davies and Rodrigo Lluberas, *Global Wealth Report 2017* (Zurich, Switzerland: Credit Suisse Research Institute, 2017), http://publications.credit-suisse.com/index.cfm/publikationen-shop/research-institute/global-wealth-report-2017-en/. Deborah Hardoon, Sophia Ayele and Ricardo Fuentes-Nieva, *An Economy for the 1%: How Privilege and Power in the Economy Drive Extreme Inequality and How This Can Be Stopped*, Oxfam Briefing Paper 210 (Oxford, UK: Oxfam, 2016), https://oxfamilibrary.openrepository.com/bitstream/handle/10546/592643/bp210-economy-one-percent-tax-havens-180116-en.pdf;jsessionidÄF4C1C3DFBF17624D6DB7F9305F0D36?sequenceG.

152 Top 100 economies: Duncan Green, "The World's Top 100 Economies: 31 Countries; 69 Corporations," *People, Spaces, Deliberation* blog (The World Bank), 20 September 2016, https://blogs.worldbank.org/publicsphere/world-s-top-100-economies-31-countries-69-corporations.

156 Leaders who seized power by coup d'etat: Clayton L. Thyne and Jonathan M. Powell, "Coup d'État or Coup d'Autocracy? How Coups Im-

pact Democratization, 1950–2008," *Foreign Policy Analysis* 12, no. 2 (2016): 192–213.

157 "Rooster Who Gets All the Hens": Ghislain C. Kabwit, "Zaire: The Roots of the Continuing Crisis," *The Journal of Modern African Studies* 17, no. 3 (1979): 381–407.

161 Structure versus agency: In social science, "agency" refers to people and the things they do. For example, a leader can make a decision that can affect a whole organization or even a nation. But social scientists don't like to call humans people. Instead of saying "individual," they use the anodyne word "agent," and an individual's influence is called agency. Agency means the power of people to control things. Conversely, "structure" refers to social structures that constrain and even control human behavior: laws, institutions, bureaucracy, norms. What social scientists call structure, everyone else calls "the establishment" or "the system." Social scientists are forever debating which is more important in human affairs, structure versus agency, when really, it's both. Social scientists are funny people—er, agents.

163 Petraeus on Iran's deep state: Jeff Schogol, "15 Years after the US Overthrew Saddam Hussein, Iraq Is Still at War with Itself," *Task and Purpose*, 19 March 2018, https://taskandpurpose.com/iraq-to-the-future.

164 *Yes, Minister* TV show: "Big Brother," *Yes, Minister*, season 1, episode 4, directed by Sydney Lotterby (London: British Broadcasting Corporation, 1980).

165 Margaret Thatcher discusses *Yes, Minister*: Michael Cockerell, *Live from Number 10: The Inside Story of Prime Ministers and Television* (London: Faber and Faber, 1988), 288.

166 Teddy Roosevelt on the US' "invisible government": Theodore Roosevelt, "Appendix B: The Control of Corporations and 'The New Freedom,'" in *Theodore Roosevelt: An Autobiography* (New York: Macmillan,

1913), published online by Bartleby (1999), www.bartleby.com/55/15b .html.

167 The "military-industrial complex": Dwight D. Eisenhower, "Farewell Radio and Television Address to the American People," The White House, United States Government, 17 January 1961, www.eisenhower .archives.gov/all_about_ike/speeches/farewell_address.pdf.

167 Deep state humor: The Eisenhower School for National Security and Resource Strategy, located at Fort Lesley J. McNair, Washington DC.

167 Senate staffer Lofgren on the deep state: Mike Lofgren, "Essay: Anatomy of the Deep State," Bill Moyers.com, 21 February 2014, http:// billmoyers.com/2014/02/21/anatomy-of-the-deep-state.

167 Scholar Glennon on "double government": Michael J. Glennon, "National Security and Double Government," *Harvard National Security Journal* 5, no. 1 (10 January 2014), available at Social Science Research Network online, https://papers.ssrn.com/sol3/papers.cfm?abstract_id =2376272&rec=1&srcabs=2323021&alg=1&pos=6.

168 Trump's reversal on Afghanistan: Donald Trump, "Remarks on the Strategy in Afghanistan and South Asia," The White House, United States Government, delivered at Fort Myer, Arlington, Virginia, 21 August 2017, www.whitehouse.gov/the-press-office/2017/08/21/remarks -president-trump-strategy-afghanistan-and-south-asia.

168 Teddy Roosevelt's solution: Roosevelt, "Appendix B."

168 Eisenhower's solution: Eisenhower, "Farewell Radio and Television Address."

RULE 8: THERE WILL BE WARS WITHOUT STATES

171 Debbie Reynolds on Acapulco: David Ehrenstein, "When Acapulco Was All the Rage: Elizabeth Taylor, Ronald Reagan and Other A-List-

ers in Mexico," *The Hollywood Reporter*, 15 March 2014, https://www.hollywoodreporter.com/news/acapulco-was-all-rage-elizabeth-687402.

171 "Today Acapulco is a battlefield": Jessica Dillinger, "The Most Dangerous Cities in the World," WorldAtlas.com, 25 April 2018, www.worldatlas.com/articles/most-dangerous-cities-in-the-world.html.

172 Evaristo on street violence: Joshua Partlow, "Acapulco Is Now Mexico's Murder Capital," *Washington Post*, 24 August 2017, www.washingtonpost.com/graphics/2017/world/how-acapulco-became-mexicos-murder-capital.

172 Sinaloa cartel: Patrick Radden Keefe, "Cocaine Incorporated," *New York Times Magazine*, 15 June 2012, www.nytimes.com/2012/06/17/magazine/how-a-mexican-drug-cartel-makes-its-billions.html.

173 The Zetas: Michael Ware, "Los Zetas Called Mexico's Most Dangerous Drug Cartel," CNN, 6 August 2009, http://edition.cnn.com/2009/WORLD/americas/08/06/mexico.drug.cartels/index.html.

174 Acapulco "like being in Afghanistan": David Agren, "Mexico after El Chapo: New Generation Fights for Control of the Cartel," *The Guardian*, 5 May 2017, www.theguardian.com/world/2017/may/05/el-chapo-sinaloa-drug-cartel-mexico.

174 Mexican police and military corruption: Jeremy Kryt, "Why the Military Will Never Beat Mexico's Cartels," *Daily Beast*, 2 April 2016, www.thedailybeast.com/why-the-military-will-never-beat-mexicos-cartels.

175 "a damning indictment" of federal forces: "Mexican Citizens Take the Drug War into Their Own Hands," *The World*, Public Radio International, produced by Christopher Woolf and Joyce Hackel, 16 January 2014, www.pri.org/stories/2014-01-16/mexican-citizens-frustrated-their-government-take-arms-against-drug-cartel.

175 US has spent $1 trillion on the War on Drugs: William R. Kelly, *The*

Future of Crime and Punishment: Smart Policies for Reducing Crime and Saving Money (Lanham, MD: Rowman & Littlefield, 2016), 50.

176 US policy makers see cartels as *West Side Story*, not war: This section is not about how to win the "War on Drugs" but rather about how modern wars do not look like wars to conventional warriors. Winning the drug wars requires a strategy that reduces supply and demand for narcotics, and this task requires more than superior firepower.

176 Mexico second deadliest country in world: The International Institute for Strategic Studies (IISS). *Armed Conflict Survey 2017* (London: Routledge/International Institute for Strategic Studies, 2017), 5.

177 Mexican cartels' GDP is $39 billion: Kelly, *Future of Crime and Punishment*, 164.

179 Battling criminal organizations is "not war": Thomas G. Mahnken, "Strategic Theory," in *Strategy in the Contemporary World: An Introduction to Strategic Studies*, 5th ed. (Oxford: Oxford University Press, 2011), 62.

180 Milton Friedman on why war: Travis Pantin, "Milton Friedman Answers Phil Donahue's Charges," *New York Sun*, 12 November 2007, www.nysun.com/business/milton-friedman-answers-phil-donahues-charges/66258.

181 "the label 'war' somehow connotes legitimacy": The line between legitimate violence in war and mass murder is blurry, if nonexistent. This book does not address directly the important ethical questions surrounding organized violence. Instead, it focuses on *how* to win wars efficiently, not *if* we should wage them. However, a better understanding of war will reduce loss of life, since strategic mistakes are paid for in blood.

181 Modern war's high civilian death rate: Kendra Dupuy et al., *Trends*

in Armed Conflict, 1046–2015 (Oslo: Peace Research Institute Oslo, August 2016), www.prio.org/utility/DownloadFile.ashx?id=15&type=publicationfile For other studies, see "COW War Data, 1817–2007 (v4.0)," The Correlates of War Project, 8 December 2011, www.correlatesofwar.org/data-sets; Meredith Reid Sarkees and Frank Wayman *Resort to War: 1816–2007* (Washington DC: CQ Press, 2010); Uppsala Conflict Data Program (UCDP)/Peace Research Institute Oslo (PRIO) Armed Conflict Dataset, 31 December 2016, www.ucdp.uu.se/gpdatabase/search.php; Nils Petter Gleditsch et al., "Armed Conflict 1946–2001: A New Dataset," *Journal of Peace Research* 39, no. 5 (2002): 615–37; Adam Roberts, "Lives and Statistics: Are 90 percent of War Victims Civilians?" *Survival* 52, no. 3 (2010): 115–36.

181 Rwandan genocide: On the 800,000 genocide deaths, see "Rwanda: How the Genocide Happened," BBC News, 17 May 2011, www.bbc.com/news/world-africa-13431486; on Iraq War deaths, see the Iraq Body Count, accessed on 6 June 2018, www.iraqbodycount.org. Per this source, 117,102 Iraqi deaths occurred in Iraq from January 2003 to December 2011.

183 Congo wars: "IRC Study Shows Congo's Neglected Crisis Leaves 5.4 Million Dead," International Rescue Committee 22 January 2008, https://reliefweb.int/report/democratic-republic-congo/irc-study-shows-congos-neglected-crisis-leaves-54-million-dead; "Body Count: Casualty Figures after 10 Years of the 'War on Terror,'" Physicians for Social Responsibility, March 2015, www.psr.org/assets/pdfs/body-count.pdf.

184 Menachem Begin wishes Iraq and Iran "good luck": Pierre Razoux, *The Iran-Iraq War,* trans. Nicholas Elliott (Cambridge, Massachusetts: Harvard University Press, 2015), 114.

186 Sacchetti story: Franco Sacchetti, *Il Trecentonovelle*, novella CLXXXI

(Torino: Einaudi, 1970), 528–29. For more on this period, see William Caferro, *John Hawkwood: An English Mercenary in Fourteenth-Century Italy* (Baltimore: JHU Press, 2006).

187 Frederick William on mercenaries during the Thirty Years' War: Sidney B. Fay, "The Beginnings of the Standing Army in Prussia," *American Historical Review* 22, no. 4 (1917): 767.

188 WWIII close calls: Superpower tensions during the Cold War that could have escalated into nuclear war include the Korean War (1950–1953); the Cuban Missile Crisis (1962); the Yom Kippur War (October 1973); NORAD's computer error (1979); the "Petrov save" incident (1983); and the Able Archer NATO exercise (1983).

RULE 9: SHADOW WARS WILL DOMINATE

197 Putin admits Russian troops in Ukraine: Shaun Walker, "Putin Admits Russian Military Presence in Ukraine for First Time," *The Guardian*, 17 December 2015, https://www.theguardian.com/world/2015/dec/17/vladimir-putin-admits-russian-military-presence-ukraine.

199 Malaysia Airlines Flight 17: Kevin G. Hall, "Russian GRU Officer Tied to 2014 Downing of Passenger Plane in Ukraine," McClatchy DC News, 25 May 2018, www.mcclatchydc.com/news/nation-world/world/article211836174.html#cardLink=row1_card2.

200 Russia's "information confrontation": Defense Intelligence Agency, *Russia Military Power: Building a Military to Support Great Power Aspirations* (2017), p. 38, www.dia.mil/Portals/27/Documents/News/Military%20Power%20Publications/Russia%20Military%20Power%20Report%202017.pdf.

202 "45,000 garbage tweets": "Russian Twitter Trolls Meddled in Brexit Vote: Did They Swing It?" *The Economist*, 27 November 2017, www

.economist.com/news/britain/21731669-evidence-so-far-suggests-only -small-campaign-new-findings-are-emerging-all.

202 US intelligence agencies on 2016 election: *Background to "Assessing Russian Activities and Intentions in Recent US Elections": The Analytic Process and Cyber Incident Attribution* (Washington: Office of the Director of National Intelligence, US Intelligence Community Assessment, declassified version, 6 January 2017), www.dni.gov/files/documents /ICA_2017_01.pdf.

202 Department of Justice charges thirteen Russians*: United States of America vs Internet Research Agency LLC, et al.*, no 1:18-cr-00032-DLF (DC, US District Court, District of Columbia, 16 February 2018), www.justice.gov/file/1035477/download.

People v Moody, No 4582-84, slip op at 3 (NY, Supreme Court, New York County, 27 June 1986).

202 USS *Maine*: The USS *Maine* blew up in Havana's harbor in 1898, and an inflamed American public blamed Spain. Yellow journalism hyped the situation, and the United States marched to war against Spain crying, "Remember the *Maine*!" Actually, the *Maine* sank due to an internal explosion and was not the work of Madrid's skullduggery.

207 Putin behind the Moscow bombings: John Dunlop, *The Moscow Bombings of September 1999: Examinations of Russian Terrorist Attacks at the Onset of Vladimir Putin's rule*, vol. 110 (Columbia University Press, 2014); Amy Knight, "Finally, We Know about the Moscow Bombings," *New York Review of Books* 22 November 2012; see also US Congress, House, Committee on Foreign Affairs, *Russia: Rebuilding the Iron Curtain, Hearing before the Committee on Foreign Affairs*, 110th Cong., 1st sess., 17 May 2007 (testimony by David Satter, senior fellow, Hudson Institute), https://web.archive.org/web/20110927065706/http://www .hudson.org/files/publications/SatterHouseTestimony2007.pdf.

209 Smedley Butler "gangster for capitalism": Smedley Butler, major general, US Marine Corps, "On Interventionism," excerpt from a speech delivered in 1933, available online at https://fas.org/man/smedley.htm.

209 John Dulles "Communist-type reign of terror": David W. Dent, ed., *US-Latin American Policymaking: A Reference Handbook* (Westport, CT, London: Greenwood Publishing Group, 1995), 81.

210 Operation PBSUCCESS: Central Intelligence Agency, *Operation PBSUCCESS: The United States and Guatemala, 1952–1954*, by Nicholas Cullather, declassified and released by the Historical Staff, Center for the Study of Intelligence, 1994, www.cia.gov/library/readingroom/docs/DOC_0000134974.pdf.

211 CIA's postmission report: Central Intelligence Agency, *Report on Stage One PBSUCCESS*, 9 December 1953, declassified and released by the Historical Staff, Center for the Study of Intelligence, 1993, www.cia.gov/library/readingroom/docs/DOC_0000928348.pdf.

214 Fake "grassroots" groups: In the private intelligence community, the practice of creating fake grassroots organizations is called "astroturfing," and it is a part of shaping operations. Regarding think tanks, Abraham Lincoln was famous for saying, "Every man has his price." So do many think tanks, which increasingly blur the line between public-policy institutes and lobbying firms. For example, see Eric Lipton et al., "Foreign Powers Buy Influence at Think Tanks," *New York Times*, 6 September 2014, www.nytimes.com/2014/09/07/us/politics/foreign-powers-buy-influence-at-think-tanks.html.

215 Moscow fears "color revolutions": Anthony H. Cordesman, "A Russian Military View of a World Destabilized by the US and the West (Full Report)," Center for Strategic and International Studies, 28 May 2014, https://csis-prod.s3.amazonaws.com/s3fs-public/legacy_files/files/publication/140529_Russia_Color_Revolution_Full.pdf.

216 "Sticky power": Sticky power is essentially economic interdependence, but weaponized. Walter Russell Mead, "America's Sticky Power," *Foreign Policy*, 29 October 2009, http://foreignpolicy.com/2009/10/29/americas-sticky-power.

217 Gaddafi son's fraudulent PhD: Rumor has it that staff at the Libyan embassy in London wrote the bulk (all?) of his dissertation while he was a student at the London School of Economics and Politics. Said Al-Islam Alqadhafi, "The Role of Civil Society in the Democratisation of Global Governance Institutions," from "'Soft Power' to Collective Decision-Making?" (thesis, London School of Economics and Politics, September 2007).

217 Top universities take bribes: Daniel Golden, "The Story Behind Jared Kushner's Curious Acceptance into Harvard," ProPublica, 18 November 2016, www.propublica.org/article/the-story-behind-jared-kushners-curious-acceptance-into-harvard.

RULE 10: VICTORY IS FUNGIBLE

221 Kühlmann's secret plan to neutralize Tsarist Russia: Zbyněk Anthony Bohuslav Zeman, ed., *Germany and the Revolution in Russia, 1915–1918: Documents from the Archives of the German Foreign Ministry* (Oxford: Oxford University Press, 1958), 193.

224 "After midnight, the siren went off": Michael Sullivan, "Recalling the Fear and Surprise of the Tet Offensive," *Morning Edition*, National Public Radio, 31 January 2008, www.npr.org/templates/story/story.php?storyId=18551391.

225 "Tet Offensive shocked Americans": On the government's credibility gap and 18 February 1968 numbers, see Clark M. Clifford and Richard C. Holbrooke, *Counsel to the President: A Memoir* (New York: Random

House, 1992), 47–55, 479. On February draft numbers and calling up the reserves, see "Johnson Considers Calling Up Reserves," United Press International, printed in the *Madera Tribune* 23 February 1968, https://cdnc.ucr.edu/cgi-bin/cdnc?a=d&d=MT19680223.2.6; on McNamara, see: Robert S. McNamara, "The Fog of War: Eleven Lessons from the Life of Robert S. McNamara," interview by Errol Morris, edited and released by Sony Pictures Classics, 2004, interview transcript available online at ErrolMorris.com, accessed on 8 June 2018, www.errolmorris.com/film/fow_transcript.html.

225 "The most trusted man in America": "Walter Cronkite Dies," *CBS Evening News with Jeff Glor*, 17 July 2009, www.cbsnews.com/news/walter-cronkite-dies.

225 "We are mired in stalemate": *Reporting Vietnam: Part One: American Journalism, 1959–1969*, ed. Milton J. Bates et al. (New York: The Library of America, 1998), 581–2.

225 "If we've lost Walter Cronkite, we've lost the country": Walter Cronkite, interview by Howard Kurtz, CNN, 28 December 2003, available online at ttp://transcripts.cnn.com/TRANSCRIPTS/0312/28/rs.00.html.

227 Bui Tin, colonel in the North Vietnamese Army: "How North Vietnam Won the War," *Asian Wall Street Journal*, 8 August 1995, p. 8.

227 "Washington dithered to the point of dereliction": Herbert R. McMaster and Jake Williams, *Dereliction of Duty: Lyndon Johnson, Robert McNamara, the Joint Chiefs of Staff, and the Lies that Led to Vietnam* (New York: HarperCollins, 1997).

227 "You know you never defeated us on the battlefield": Harry G. Summers, *On Strategy: A Critical Analysis of the Vietnam War* (New York: Presidio Press, 1995), 1.

228 Ben Franklin, propagandist: Hugh T. Harrington, "Propaganda Warfare: Benjamin Franklin Fakes a Newspaper," *Journal of the American*

Revolution, 10 November 2014, https://allthingsliberty.com/2014/11/propaganda-warfare-benjamin-franklin-fakes-a-newspaper.

228 "the printing press is the greatest weapon": Thomas Edward Lawrence, *Evolution of a Revolt*, originally published in *Army Quarterly* 1, no. 1 (October 1920), published in the United States by Praetorian Press in 2011.

228 "Irregular war [is] far more intellectual than a bayonet charge": Thomas Edward Lawrence, *Seven Pillars of Wisdom* (privately printed, 1926; reprint, New York: Penguin Books, 1962), 348.

229 General Giap on why America lost Vietnam: Neil Sheehan, "David and Goliath in Vietnam," *New York Times*, 26 May 2017, www.nytimes.com/2017/05/26/opinion/sunday/david-and-goliath-in-vietnam.html.

229 "The guerrilla wins if he does not lose": Henry Kissinger, "The Vietnam Negotiations," *Foreign Affairs* 48, no. 2 (January 1969): 214.

233 "Tactics without strategy is the noise before defeat": This quote has long been attributed to Sun Tzu, but it does not appear in his book *The Art of War*.

235 "Gettysburg was a tactical event, not a strategic one": Gettysburg led to a strategic outcome but did not independently cause it. Many factors produced the North's victory over the South, from industrialization to political leadership to sheer numbers to the South's offensive campaign northward, which depleted its resources. Tactics are not strategy, even if they lead to strategic outcomes.

237 Hybrid war: Frank Hoffman, *Conflict in the 21st Century: The Rise of Hybrid Wars* (Arlington: Potomac Institute for Policy Studies, 2007).

239 Dunford's commencement speech at the National Defense University: Joseph Dunford, general, US Marine Corps, remarks at National Defense University graduation ceremony, Fort McNair, Washington, DC, 10 June 2016, available at www.jcs.mil/Media/Speeches/Article

/797847/gen-dunfords-remarks-at-the-national-defense-university-graduation.

240 "All things are ready, if our mind be so": William Shakespeare, *Henry V*, act 4, scene 3. Based on an actual event, the St Crispin's Day speech was delivered on October 25, 1415 to rouse a badly outnumbered English army before the Battle of Agincourt. Shakespeare reveals a secret every good warrior knows: the mind is the most powerful weapon we possess.

WINNING THE FUTURE

244 Israel's "Campaign between Wars": "Deterring Terror: How Israel Confronts the Next Generation of Threats," English translation of the *Official Strategy of the Israel Defense Forces*, Harvard Kennedy School Belfer Center for Science and International Affairs, Special Report, August 2016, www.belfercenter.org/sites/default/files/legacy/files/IDFDoctrine Translation.pdf.

245 Russia cut military spending 20 percent: David Brennan, "Why Is Russia Cutting Military Spending?," *Newsweek*, 8 June 2018, www.news week.com/why-russia-cutting-military-spending-908069.

ANNEX: THE THIRTY-SIX ANCIENT CHINESE STRATAGEMS FOR WAR

253 The Thirty-Six Stratagems: The material presented here is based on Harro von Senger, *The Book of Stratagems: Tactics for Triumph and Survival*, ed. and trans. Myron B. Gubitz (New York: Penguin Books, 1993), 369–70.

SELECTED BIBLIOGRAPHY

Arreguin-Toft, Ivan. "How the Weak Win Wars: A Theory of Asymmetric Conflict." *International Security* 26, no. 1 (2001): 93–128.

Bacevich, Andrew J. *The New American Militarism: How Americans Are Seduced by War*. New York: Oxford University Press, 2013.

Bagehot, Walter. *The English Constitution*. London: Oxford University Press, 1956.

Barylski, Robert V. *The Soldier in Russian Politics: Duty, Dictatorship and Democracy Under Gorbachev and Yeltsin*. New Brunswick, NJ: Transaction Publishers, 1998.

Betts, Richard K. "Is Strategy an Illusion?" *International Security* 25, no. 2 (Fall 2000): 5–50.

Bicheno, Hugh. *Vendetta: High Art and Low Cunning at the Birth of the Renaissance*. London: Weidenfeld & Nicolson, 2008.

Biddle, Stephen. "Strategy in War." *Political Science & Politics* 40, no. 3 (2007): 461–66.

Boot, Max. *The Road Not Taken: Edward Lansdale and the American Tragedy in Vietnam*. New York: Liveright Publishing Corporation, 2018.

Brooks, Rosa. *How Everything Became War and the Military Became Everything: Tales from the Pentagon*. New York: Simon & Schuster, 2017.

Caferro, William. *Contesting the Renaissance*. Malden, MA: Wiley-Blackwell, 2011.

Callwell, Charles Edward. *Small Wars: Their Principles and Practice*. Wakefield, UK: EP Publishing, 1976.

Card, Orson Scott. *Ender's Game*. New York: Tor, 1985.

Cheng, Dean. "Winning without Fighting: Chinese Legal Warfare." *Heritage Foundation Backgrounder* no. 2692. May 2012.

Clausewitz, Carl von. *On War*. Translated by Michael Howard and Peter Paret. Princeton, NJ: Princeton University Press, 1976.

Cohen, Eliot. *Supreme Command: Soldiers, Statesmen, and Leaders in Wartime*. New York: Anchor Books, 2002.

——— and John Gooch. *Military Misfortunes: The Anatomy of Failure in War*. New York: Free Press, 1990.

Coker, Christopher. *Future War*. Malden, MA: Polity Press, 2015.

———. *Rebooting Clausewitz: 'On War' in the Twenty-First Century*. New York: Oxford University Press, 2017.

Crowl, Philip A. "The Strategist's Short Catechism: Six Questions Without Answers." Harmon Memorial Lectures in Military History, no. 20. Air Force Academy, CO: United States Air Force Academy, 1977.

Dörner, Dietrich. *The Logic of Failure : Recognizing and Avoiding Error in Complex Situations*. Reading, Mass.: Perseus Books, 1996.

Drezner, Daniel W. *The Ideas Industry*. New York: Oxford University Press, 2017.

Echevarria II, Antulio J. *Reconsidering the American Way of War: US Military Practice from the Revolution to Afghanistan*. Washington: Georgetown University Press, 2014.

Evans, Michael. "Sun Tzu and the Chinese Military Mind." *Quadrant* 55, no. 9 (2011): 62–68.

Freedman, Lawrence. *Strategy: A History*. New York: Oxford University Press, 2013.

———. *The Future of War: A History*. New York: Public Affairs, 2017.

Galeotti, Mark. *Global Crime Today: The Changing Face of Organized Crime*. New York: Routledge, 2014.

Galula, David. *Counterinsurgency Warfare: Theory and Practice*. Westport, CT: Praeger Security International, 2006.

Gentile, Gian. *Wrong Turn: America's Deadly Embrace of Counterinsurgency*. New York: New Press, 2013.

Gerasimov, Valery. "The Value of Science Is in the Foresight: New Challenges Demand Rethinking the Forms and Methods of Carrying Out Combat Operations." Translated by Robert Coalson. *Military Review*, January–February 2016, p. 23–29.

Gray, Colin S. "Why Strategy Is Difficult." *Joint Force Quarterly* 34 (Spring 2003): 80–86.

———. *Fighting Talk: Forty Maxims on War, Peace, and Strategy*. Westport, CT: Praeger Security International, 2007.

Grayling, A. C. *War: An Enquiry*. New Haven: Yale University, 2017.

Guevara, Ernesto. *On Guerrilla Warfare*. Westport, CT: Praeger, 1961.

Hammes, Thomas X. *The Sling and the Stone: On War in the 21st Century*. Minneapolis: Zenith Press, 2006.

Hart, Liddell. *Strategy: The Indirect Approach*. New York: Faber, 1967.

Hill, Charles. *Grand Strategies: Literature, Statecraft, and World Order*. New Haven: Yale University Press, 2010.

Hoffman, Frank G. *Conflict in the 21st Century: The Rise of Hybrid Wars*. Arlington, VA: Potomac Institute for Policy Studies, 2007.

Howard, Michael. "Grand Strategy in the 20th Century," *Defence Studies* 1, no. 1 (Spring 2001): 1–10.

———. *War in European History*. New York: Oxford University Press, 2009.

Johnson, Robert. "Hard Truths, Uncomfortable Futures and the Perils of Group-Think: Rethinking Defence in the Twenty-First Century." Unpublished paper, Changing Character of War Centre, University of Oxford, 2017.

Jones, Milo, and Philippe Silberzahn. *Constructing Cassandra: Reframing Intelligence Failure at the CIA, 1947–2001*. Stanford, CA: Stanford University Press, 2013.

Josephus, Flavius. *The Jewish War*. Translated by G. A. Williamson. New York: Penguin Books, 1970.

Jullien, François. *A Treatise on Efficacy : Between Western and Chinese Thinking*. Honolulu: University of Hawai'i Press, 2004.

Kane, Thomas M. *Ancient China and Postmodern War: Enduring Ideas from the Chinese Strategic Tradition*. London: Routledge, 2007.

Kautilya, *The Arthashastra*. Translated by L. N. Rangarajan. New Delhi: Penguin, 1992.

Kennedy, Paul. "Grand Strategy in War and Peace: Toward a Broader Definition," in *Grand Strategies in War and Peace*, edited by Paul Kennedy. New Haven, CT: Yale University Press, 1991.

Krasner, Stephen D. "Abiding Sovereignty." *International Political Science Review* 22, no. 3 (2001): 229–51.

Lang, Michael. "Globalization and Its History." *Journal of Modern History* 78, no. 4 (2006): 899–931.

Lawrence, Thomas Edward. "Guerrilla Warfare." Entry in *Encyclopedia Britannica: A New Survey of Universal Knowledge*. London: Encyclopaedia Britannica Co., 1929.

———. "The Evolution of a Revolt." *Army Quarterly and Defence Journal*, October 1920, pp. 55–69.

———. *Seven Pillars of Wisdom*. New York: Penguin Books, 1962.

Liang, Qiao, and Wang Xiangsui. *Unrestricted Warfare.* Beijing: PLA Literature and Arts Publishing House, 1999.

Machiavelli, Niccolò. *The Portable Machiavelli.* Edited by Peter E. Bondanella and Mark Musa. New York: Penguin Books, 1979.

Mack, Andrew. "Why Big Nations Lose Small Wars: The Politics of Asymmetric Conflict." *World Politics* 27, no. 2 (1975): 175–200.

Mackubin, Thomas Owens. "Strategy and the Strategic Way of Thinking." *Naval War College Review* 60, no. 4 (2007): 111–24.

Mallett, Michael. *Mercenaries and Their Masters: Warfare in Renaissance Italy.* London: The Bodley Head, 1974.

——— and Christine Shaw. *The Italian Wars, 1494-1559: War, State and Society in Early Modern Europe.* London: Routledge, 2012.

Mao, Zedong. *On Guerrilla Warfare.* New York: Praeger, 1961.

———. *On Protracted War.* Beijing: Foreign Languages Press, 1966.

Marks, Thomas A. *Maoist Insurgency since Vietnam.* New York: Routledge, 2012.

McMaster, H. R. and Jake Williams. *Dereliction of Duty: Lyndon Johnson, Robert McNamara, the Joint Chiefs of Staff, and the Lies That Led to Vietnam.* New York: HarperPerennial, 1997.

Mead, Walter Russell. "America's Sticky Power." *Foreign Policy*, 29 October 2009.

Monaghan, Andrew. "The 'War' in Russia's 'Hybrid Warfare.'" *Parameters* 45, no. 4 (2015): 65.

Murray, Williamson and Mark Grimsley. "Introduction: On Strategy." in *The Making of Strategy: Rulers, States, and War*, edited by Williamson Murray, MacGregor Knox, and Alvin Bernstein. New York: Cambridge University Press, 1994.

Musashi, Miyamoto. *The Complete Book of Five Rings.* Translated by Kenji Tokitsu. Boston: Shambhala, 2010.

Olson, William. J. "The Continuing Irrelevance of Clausewitz." *Small Wars Journal*, July 2013. http://indianstrategicknowledgeonline.com/web/Small%20Wars%20Journal%20-%20The%20Continuing%20Irrelevance%20of%20Clausewitz%20-%202013-07-26.pdf.

———. "Global Revolution and the American Dilemma," *Strategic Review*, Spring 1983, pp. 48–53.

———. "The Natural Law of Strategy: A Contrarian's Lament." *Small Wars Journal*, August 2011. http://smallwarsjournal.com/jrnl/art/the-natural-law-of-strategy.

———. "The Slow Motion Coup: Militarization and the Implications of Eisenhower's Prescience." *Small Wars Journal*, August 2012. http://smallwarsjournal.com/jrnl/art/the-slow-motion-coup-militarization-and-the-implications-of-eisenhower's-prescience.

———. "War Without a Center of Gravity: Reflections on Terrorism and Post-Modern War." *Small Wars and Insurgencies*, April 2007, pp. 559–583.

Osiander, Andreas. "Sovereignty, International Relations, and the Westphalian Myth." *International Organization* 55, no. 2 (2001): 251–87.

Paret, Peter, Gordon Alexander Craig, and Felix Gilbert. *Makers of Modern Strategy: From Machiavelli to the Nuclear Age*. Princeton, NJ: Princeton University Press, 1986.

Parker, Geoffrey. *The Military Revolution: Military Innovation and the Rise of the West, 1500–1800*. Cambridge, UK: Cambridge University Press, 1988.

Parrott, David. *The Business of War: Military Enterprise and Military Revolution in Early Modern Europe*. Cambridge, UK: Cambridge University Press, 2012.

Patterson, Malcolm Hugh. *Privatising Peace: A Corporate Adjunct to United Nations Peacekeeping and Humanitarian Operations*. New York: Palgrave Macmillan, 2009.

Phillips, Thomas R. *Roots of Strategy: The 5 Greatest Military Classics of All Time*. Harrisburg, PA: Stackpole Books; 1985.

Porch, Douglas. *Counterinsurgency: Exposing the Myths of the New Way of War*. New York: Cambridge University Press, 2013.

Ricks, Thomas E. *The Generals: American Military Command from World War II to Today*. New York: Penguin, 2012.

Roberts, Michael. *The Military Revolution, 1560–1660*. Belfast: Boyd, 1956.

Rogers, Clifford J. *The Military Revolution Debate: Readings on the Military Transformation of Early Modern Europe*. Boulder: Westview Press, 1995.

Rotberg, Robert, ed. *When States Fail: Causes and Consequences*. Princeton, NJ: Princeton University Press, 2004.

Rumelt, Richard P. *Good Strategy/Bad Strategy: The Difference and Why It Matters*. New York, NY: Crown Business, 2011.

Sawyer, R. D., and M. Sawyer. *The Seven Military Classics of Ancient China*. New York: Basic Books, 2007.

Senger, Harro von. *The Book of Stratagems: Tactics for Triumph and Survival*. Edited by Myron B. Gubitz. New York: Viking, 1991.

Simpson, Emile. *War From the Ground Up: Twenty-First Century Combat as Politics*. New York: Columbia University Press, 2012.

Smith, Rupert. *The Utility of Force: The Art of War in the Modern World*. London: Allen Lane, 2005.

Strachan, Hew. *The Direction of War*. New York: Cambridge University Press, 2012.

Strange, Susan. *The Retreat of the State: The Diffusion of Power in the World Economy*. New York: Cambridge University Press, 1996.

Sun Tzu. *The Art of War*. Translated by Thomas Cleary. Boston: Shambhala, 2005.

———. *The Art of War*. Translated by Victor H. Mair. New York: Columbia University Press, 2007.

———. *The Art of War.* Translated by John Minford. New York: Penguin Press, 2002.

Taber, R. *The War of the Flea: A Study of Guerrilla Warfare Theory and Practice.* New York: Citadel Press, 1970.

Takuan Sōhō. *The Unfettered Mind: Writings of the Zen Master to the Sword Master.* Translated by William Scott Wilson. Boulder: Shambhala, 2012.

Thomson, Janice E. *Mercenaries, Pirates, and Sovereigns.* Princeton, NJ: Princeton University Press, 1996.

Tierney, Dominic. *The Right Way to Lose a War: America in an Age of Unwinnable Conflicts.* New York: Little, Brown and Company, 2015.

Tilly, Charles. "War Making and State Making as Organized Crime," in *Bringing the State Back In*, edited by Peter B. Evans, Dietrich Rueschmeyer, and Theda Skocpol. Cambridge, UK: Cambridge University Press, 1985.

Toft, Monica Duffy. *Securing the Peace: The Durable Settlement of Civil Wars.* Princeton, NJ: Princeton University Press, 2009.

Trinquier, Roger. *Modern Warfare: A French View of Counterinsurgency.* Westport, CT: Praeger Security International, 2006.

Tsunetomo,Yamamoto. *Hagakure: The Secret Wisdom of the Samurai.* Translated by Alexander Bennett. North Clarendon: Tuttle Publishing, 2014.

US Department of Defense. Joint Publication 1. *Doctrine for the Armed Forces of the United States.* Washington, DC, 2017.

———. Joint Publication 3-0. *Joint Operations.* Washington, DC, 2017.

———. *Summary of the 2018 National Defense Strategy of the United States of America: Sharpening the American Military's Competitive Edge.* Washington, DC, 2018.

US Department of the Army. Field Manual 3-24. *Counterinsurgency.* Washington, DC, 2006.

US Marine Corps. Fleet Marine Force Reference Publication 12-15. *Small Wars Manual.* Washington, DC, 1990.

Valentino, Benjamin, Paul Huth, and Dylan Balch-Lindsay. " 'Draining the Sea': Mass Killing and Guerrilla Warfare." *International Organization* 58, no. 2 (2004): 375–407.

Vlahos, Michael. *Fighting Identity: Sacred War and World Change.* Westport, CT: Praeger, 2008.

Weigley, Russell Frank. *The American Way of War: A History of United States Military Strategy and Policy.* Bloomington: Indiana University Press, 1977.

Wilson, Peter H. *The Thirty Years War: Europe's Tragedy.* Cambridge, Mass.: Harvard University Press, 2009.

Woodmansee, John W. "Mao's Protracted War: Theory vs. Practice." *Parameters* 3, no. 1 (1973): 30–45.

Wright, Thomas. *All Measures Short of War: The Contest for the Twenty-First Century and the Future of American Power.* New Haven, CT: Yale University Press, 2017.

INDEX